Cool Energy

Cool Energy

Renewable Solutions to Environmental Problems

revised edition

Michael Brower

The MIT Press
Cambridge, Massachusetts
London, England

Second printing, 1993

This book was printed and bound in the United States of America.

Library of Congress Cataloging-in-Publication Data

Brower, Michael, 1960 –
Cool energy: renewable solutions to environmental problems/
Michael Brower. — Rev. ed.
p. cm.
Includes bibliographical references and index.
ISBN 0-262-02349-0. — ISBN 0-262-52175-X (pbk.)
1. Renewable energy sources — United States. 2. Environmental protection—United States. I. Title.
TJ807.9.U6B76 1992
333.79'4'0973 — dc20 92-17472
CIP

Contents

Acknowledgments vii
Introduction 1

1 The Energy Challenge 5

2 The Renewable Alternative 21

3 Solar Energy 39

4 Wind Energy 71

5 Biomass 87

6 Energy from Rivers and Oceans 111

7 Geothermal Energy 127

8 Energy Storage 155

9 Policies for a Renewable Future 173

Appendix A: Units and Conversion Factors 187
Appendix B: U.S. Renewable Energy Funding 190
Suggested Readings 193
References 197
Index 217

Acknowledgments

Many people provided invaluable help to me in researching and writing this report. I am especially indebted to the reviewers: Donald Aitken (Union of Concerned Scientists), David Anderson (Geothermal Resources Council), Michael Bergey (Bergey Windpower), James Birk (Electric Power Research Institute), Eldon Boes (National Renewable Energy Laboratory), Robert Cohen, Janet Cushman (Oak Ridge National Laboratory), David Duchane (Los Alamos National Laboratory), Helen English (Passive Solar Industries Council), Paul Gipe, Paul Lienau (Geo-Heat Center), Michael Lotker (Siemens Solar), Jane Negus-deWys (Idaho National Engineering Laboratory), Kevin Porter (National Renewable Energy Laboratory), Randall Swisher (American Wind Energy Association), Robert Thresher (National Renewable Energy Laboratory), and Shaine Tyson (National Renewable Energy Laboratory). Although they did their utmost to correct errors of fact and judgment in the manuscript, any that remain are, of course, my responsibility.

This work was very much a collaborative effort with my colleagues at the Union of Concerned Scientists. Patrick Skerrett, Levan Hiemke, and Eric Denzler provided valuable assistance in researching and writing several portions. I am especially grateful to Warren Leon, Herb Rich, and Jan Wager for endless hours spent editing, proofreading, and producing the book.

Research and writing of this study were supported in part by grants from the Energy Foundation, Joyce Mertz-Gilmore Foundation, Public Welfare Foundation, and Florence and John Schumann Foundation.

M.C.B.

Cool Energy

Introduction

The United States is rapidly approaching a new energy crisis. Consumption of energy is growing, but at the same time, Americans are increasingly demanding that the energy come from clean, safe sources that protect their health, environment, and quality of life. This sharpening conflict is seen in the virtual halt to nuclear power plant construction, strict new federal and state controls on pollution from fossil-fueled power plants, automobiles, and industry, and the high-profile debate over what to do about global warming.

Renewable energy sources such as wind, sunlight, plants, and geothermal energy could provide a way out of this emerging crisis. Although many in government and more than a few other Americans expect our nation's dependence on oil, coal, and natural gas to extend indefinitely into the future, much of the technology has already been developed to allow us to move in a different direction. This book makes the case that it is not only desirable but practical to make the transition from fossil fuels to "cool" renewable energy. By doing so, we will help preserve the environment and sustain our economy at the same time.

Renewable energy — that which is regenerated at the same rate it is used — was first widely considered as an alternative to fossil fuels in the seventies. In response to the decade's oil crises, it enjoyed a brief period of popularity. Under President Jimmy Carter's administration, funding for research and development in this area grew from almost nothing to more than $700 million in 1980. Tax credits and other programs made solar collectors, wind turbines, and other devices attractive business investments. So many families — from the Carters on down — placed solar collectors on the roofs of their houses that collector sales increased fivefold between 1975 and 1980.

Unfortunately, falling oil prices in the eighties put renewable energy on the back burner. A new administration under Ronald Reagan, deeply hostile to anything other than the energy status quo, drastically reduced funding for research and even removed the solar collectors from the White House roof.

Now, in the nineties, when there is greater awareness of the environmental problems caused by excessive dependence on fossil fuels, government officials, business leaders, and the media remain skeptical that renewable energy technologies can be deployed on a large enough scale to displace significant quantities of oil, coal, or natural gas. The brief boom in renewable energy did not last long enough to convince people that the future lies in wide-scale use of solar, wind, biofuels, or geothermal energy. In fact, because some of the crash implementation projects of the late seventies were poorly conceived — and in some cases the technologies simply were not ready for deployment — renewable energy earned a reputation for high cost and unreliability.

Renewable energy technologies nevertheless made dramatic strides while they were out of the public eye. The reliability and efficiency of equipment improved; the cost of installing, maintaining, and running it declined. Moreover, energy planners gained a better appreciation of how these technologies could be integrated efficiently and reliably into the existing energy system.

In the case of wind turbines, for example, more advanced designs, better choice of materials, and careful siting have made the cost of generating electricity from wind a fourth of what it was a decade ago. In many locations, a utility company can now build a wind-power facility that will produce electricity at a cost approaching that of a new fossil-fuel power plant. And, if such hidden costs of fossil fuels as air pollution and global warming are considered, wind can be a cheaper source of electricity than fossil fuels.

This book begins with a chapter describing some of the economic and environmental consequences of America's fossil-fuel-based economy. It makes the case that, despite some progress in reducing pollution from fossil fuels, no lasting cure for our deteriorating environment — in particular, the looming threat of global warming — is possible without developing alternative fuel sources. Just as important, America's economic security is becoming increasingly vulnerable to the actions of just a few oil-producing states, as the 1991 Persian Gulf war demonstrated. Reducing fossil-fuel use thus makes both environmental and economic sense.

That renewable energy can provide the bulk of the new supplies we need is the theme of the second chapter, which discusses the relative advantages of these resources compared to fossil fuels and nuclear power and evaluates their long-term potential. Yet the chapter also reveals that progress in commercializing renewable energy will be extremely slow unless the government acts to remove a number of market barriers. These barriers include, for example, a tax system that unfairly penalizes investments in renewable energy, utility regulations that encourage continued reliance on fossil fuels, and the failure of markets to account for the long-term economic, environmental, and social consequences of energy choices.

The bulk of the book considers five broad categories of renewable energy sources: solar, wind, biomass (plant matter), rivers and oceans, and geothermal. For each of these sources, the book describes its current application, discusses its costs, analyzes new technologies under development, and assesses its positive and negative environmental impacts. Because conventional wisdom holds that inherently fluctuating renewable resources like solar and wind cannot make significant inroads into world supply without energy storage, the book devotes a chapter to the energy storage issue. It concludes with a chapter on policies that could help speed the transition to a renewable energy economy.

Taken as a whole, this book shows the vital role renewable sources can and should play in America's energy future. It cites studies indicating that, with the right policies, renewable energy could provide as much as half of America's energy within 40 years, and an even larger fraction down the road. Such a rapid shift from existing energy sources would be dramatic but not unprecedented. In 1920, coal supplied 70 percent of U.S. energy, but within 40 years its share had dropped to just 20 percent as oil and natural gas use increased.

Sooner or later, oil and natural gas will also fade in importance. The real question is when. This book makes the case that the time to move decisively toward a renewable energy economy has arrived.

1 The Energy Challenge

From almost the start of the industrial revolution, the engine of Western civilization has run on fossil fuels. In the United States, coal emerged as the dominant energy source several decades after it did in Europe, but by the end of the 19th century it was providing fully 50 percent of U.S. energy needs. Oil and natural gas came into wide use in the 20th century, helping to reduce the cost and improve the quality of transportation, industry, residential heating, and other energy services. Today, close to 85 percent of U.S. energy needs are met in one manner or another by fossil fuels.

Yet this dependence cannot continue without putting the U.S. economy and the global environment at risk. Even before the threat of global warming attracted wide attention in the late eighties, it was clear that U.S. and world oil and natural gas reserves would not last indefinitely, and that prudence called for gradually reducing our dependence on these energy sources. At present rates of consumption, proven U.S. oil reserves will last just 10 years, and world oil reserves will last barely 40 years. To be sure, new reserves are being discovered all the time, but even if total reserves prove to be double current estimates, the world will begin running short on oil and natural gas by the middle of the next century. Well before then, prices are likely to rise sharply because of the increasing difficulty and cost of discovering and extracting oil. The growing concentration of oil production in a small number of countries is also cause for alarm. By 2020, if present trends continue, over two-thirds of world oil will be pumped from the Middle East, compared to just a quarter today — a deeply troubling prospect, considering the instability and conflict that continue to plague that region (Flavin and Lenssen 1990).

Given these trends, the world seems destined to experience another period of sharply rising energy prices like that which struck in the seventies, this one possibly far more severe and enduring than the last.

The potential impact of the next oil crisis on the U.S. economy cannot be overstated. About 40 percent of U.S. energy comes from oil. Even though half of this is domestically produced (a fraction certain to fall over the next several years as domestic reserves are depleted), the oil market knows no national boundaries. If the world oil price rises, then all of the oil we consume will become more expensive. Even more distressing is the possibility of a complete cutoff of Persian Gulf oil, resulting perhaps from war, which would shut down much of American industry and cause crippling shortages of gasoline and heating oil.

The Greenhouse Connection

All of these dangers were widely recognized in the seventies. What is new today is the growing appreciation by both scientists and the public at large of the role fossil fuels play in damaging the global environment. To be sure, Americans have long been troubled by air pollution, which continues to affect the health and comfort of millions of city dwellers. Even our rural and wilderness areas are not immune to this problem: High levels of ozone (a component of smog) are damaging crops and reducing agricultural productivity, while acid rain and clouds are strongly suspected of damaging forests and poisoning lakes and streams in parts of the eastern United States and Canada (WRI 1990, MacKenzie and El-Ashry 1988).

Until very recently, however, such problems were believed to be largely local, or at worst regional, phenomena. For the public, the first inkling of the error in this thinking came when scientists announced discovery, in 1986, of the now-famous "hole" in the stratospheric ozone layer over Antarctica, which vividly demonstrated that localized emissions of pollutants could cause global environmental damage. This hole is caused not by fossil-fuel combustion but by the release of a class of chemicals known as chlorofluorocarbons, or CFCs, long-lived substances used in refrigerators, air conditioners, solvents, and other applications. Stratospheric ozone — not to be confused with tropospheric, or near-ground-level, ozone — blocks the sun's harmful ultraviolet rays from reaching the earth's surface. Consequently, a decrease in stratospheric ozone results in an increase in the exposure of people to ultraviolet radiation, and thus very likely an increase in the incidence of afflictions such as skin cancer and cataracts. The destruction of ozone is now observed to be spreading to temperate regions and has become the object

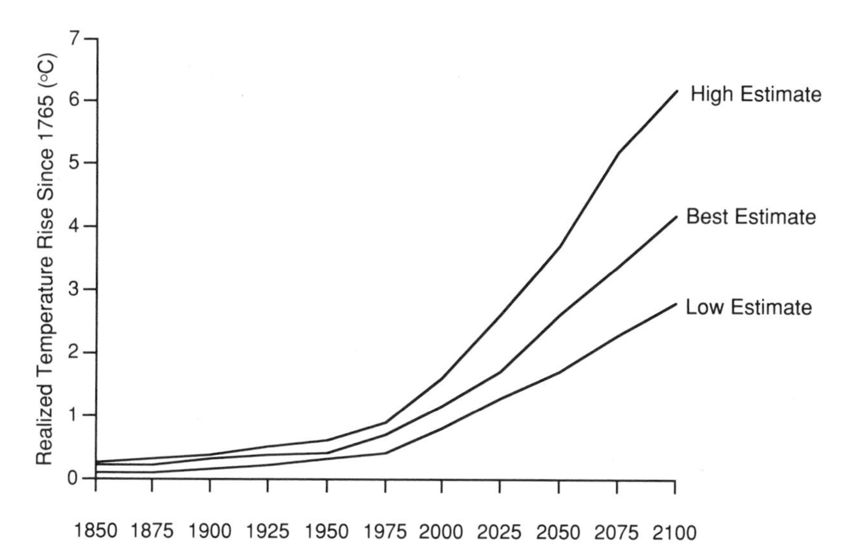

Figure 1.1
Computer models of the earth's climate indicate that the global average temperature has risen 0.5°C – 1.5°C since 1765, and that it is likely to rise an additional 2.5°C – 4.5°C by the end of the next century, because of emissions of greenhouse gases (IPCC 1990).

of international agreements to reduce and eventually eliminate production of CFCs.

More recently, attention has been focused on the threat of global warming. Scientists have long known that certain gases in the atmosphere absorb heat (infrared light) radiating from the earth's surface. Most of these gases, such as water vapor, carbon dioxide, and methane, exist naturally, and without their warming influence — the greenhouse effect — the earth would be much colder and uninhabitable. Various human activities are believed to be aggravating the greenhouse effect, however, by adding ever-greater quantities of these gases to the atmosphere. From direct measurements since 1958, and before then from gas samples taken from ice cores, we know that the atmospheric concentration of carbon dioxide is up 25 percent since preindustrial times. The concentrations of other greenhouse gases such as methane and CFCs[1] are also rising (IPCC 1990, Bolin et al. 1986, Ramanathan 1988).

Global warming first came to the public's attention in the summer of 1988, when droughts and heat waves gripped much of the United States causing crops to fail and rivers to dry up. In a dramatic announcement, a leading climatologist, James Hansen, told a Congressional committee he was "98 percent" sure that a significant global temperature rise was occurring, although whether this was directly the result of accumulating greenhouse gases he could not say. Indeed, the temperature trend is alarming: An increase of 0.3°C to 0.6°C (0.5°F to 1.0°F) in the earth's average temperature has been observed in meteorological records going back over 100 years. What is most striking is that the seven warmest years on record have all occurred since 1980. These facts cannot be ascribed unambiguously to an artificial increase in the greenhouse effect, however, since many other factors (such as variations in solar radiation and changes in ocean currents) can also influence climate.

Nevertheless, there is a strong consensus among scientists that the earth is likely to get warmer in the future. A recent international study sponsored by several governments (including the United States) and involving some 300 scientists from various disciplines concluded the following:

> We are certain [that] there is a natural greenhouse effect which already keeps the Earth warmer than it would otherwise be [and that] emissions resulting from human activities are substantially increasing the atmospheric concentrations of the greenhouse gases. . . .These increases will enhance the greenhouse effect, resulting on average in an additional warming of the Earth's surface. (IPCC 1990)

If present trends continue, the concentrations of all greenhouse gases will rise to the equivalent of double the preindustrial concentration of carbon dioxide by around 2020. According to the best climate models available, such a rise will cause the earth to warm anywhere from 2.5°C to 4.5°C (4.5°F to 8.1°F) by the end of the next century. (The earth would actually be *committed* to a greater increase than that, but the warming would be delayed several decades by the thermal inertia of the oceans.)

Predicting future warming with any accuracy is difficult because many complex climate-related processes are either poorly understood or difficult to simulate with existing computer capabilities. For example, scientists cannot identify or quantify all of the natural sources and "sinks" of carbon dioxide and so cannot predict accurately how much carbon-dioxide levels will increase in the future, even if human carbon-dioxide emissions could be predicted perfectly. There are also significant scientific uncertainties concerning cloud formation and dissipation, the exchange of energy between the oceans and the atmosphere, and the behavior of polar ice sheets, all of which have an important impact on warming predictions. In addition, today's computer models do not have the capacity to predict small-scale or regional climate variations with any accuracy.

Yet if the model predictions hold true, the temperature increase will be more sustained and rapid than anything modern humans have ever experienced. To put it in perspective, today's global average temperature is only about 5°C (9°F) warmer than it was at the peak of the last Ice Age 18,000 years ago, when much of the Northern Hemisphere was covered by ice sheets kilometers thick (Schneider and Londer 1984). The effects of an equally large — and far more rapid — future warming would be profound and irreversible, and very likely adverse to the human race.

In 1988 the Environmental Protection Agency released a study of the potential impacts of global warming on the United States. It suggested, among other things, that forest systems would begin to decline within a few decades and a number of species of plants and animals would die out because they could not migrate north quickly enough or their paths of migration would be blocked by urban sprawl. Most of the country's coastal wetlands — many of them irreplaceable wildlife refuges — would be lost to rising seas (caused by the melting of land-based polar ice and the thermal expansion of seawater). Coastal communities would have to spend large sums to protect against flooding. Agricultural output would be affected, as forecasters predict increased summer dryness in the American breadbasket and a higher frequency of droughts and heat waves (although increased levels of atmospheric carbon dioxide could at least partially offset these changes by aiding plant growth). In some parts of the country, water for drinking, irrigation, and industry would become more scarce (EPA 1988).

Striking a somewhat more optimistic note, a 1991 National Academy of Sciences study concluded that because of its rich natural and human resources, the United States "is well situated to respond to greenhouse warming," although the cost would be high (NAS 1991). Yet even if the United States can adapt to global warming without extreme disruptions, its effects could well take on tragic proportions in other parts of the world, particularly in less-developed countries ill-equipped to cope with rapidly changing climate conditions. Famines could occur as heavily populated, food-producing coastal regions are inundated by rising seas and the interiors of continents are afflicted by more frequent droughts. Increased stress on natural ecosystems could lead to mass refugee movements and possibly even wars over scarce resources. Considering the ever-growing interdependency of world economies, it seems unlikely that the United States could remain wholly removed from such disasters.

Taken together, the possible effects of global warming present a frightening threat to future generations. Some of the effects, moreover, may not be anticipated, just as scientists failed to anticipate the Antarctic ozone hole. Most studies assume, for example, that whatever climate changes occur will be gradual, but there are indications — in the cyclical history of ice ages, for example — that the earth's climate can change abruptly and radically in response to unknown factors. As the 1988 EPA study noted, predictions are "inherently limited by our imaginations. . . . Until a severe event occurs . . . we fail to recognize the close links between our society, the environment, and climate."

The Role of Fossil Fuels

A variety of human activities is contributing to the release of greenhouse gases into the atmosphere. They include the destruction of tropical rainforests and the associated release of carbon dioxide, methane, and other gases; agricultural practices, such as the use of nitrogen-rich fertilizers, which generates nitrous oxide, and the growing of rice in flooded paddies, which produces methane; and emissions of CFCs.

The chief source of greenhouse gases, however, is the combustion of fossil fuels. Worldwide, fossil-fuel combustion accounts for more than 70 percent of all human carbon-dioxide emissions — approximately 20 billion metric tons annually — and carbon dioxide accounts for 55 percent of world contributions to global warming (IPCC 1990, Houghton and Woodwell 1989). Fossil fuels are also a source of nitrous oxide, methane, and, indirectly, tropospheric (low-altitude) ozone, which is not only a pollutant but a greenhouse gas. Nitrous oxide is produced in combustion, methane through leaks from natural-gas wells, pipelines, and coal mines, and ozone through photochemical reactions involving methane, nitrogen oxides, and other compounds. All told, fossil-fuel use accounts for about half of the warming that is estimated to have occurred in the 1980s and over half of the warming predicted for the next 100 years (EPA 1990).

All countries contribute to global warming to some degree, but the United States bears an especially heavy responsibility. Although it has just 5 percent of the world's population, the United States contributes about 24 percent of world carbon-dioxide emissions from fossil fuels. When the cumulative effects of past emissions are considered, the U.S. contribution is even greater—30 percent for the period 1950 to 1987 (WRI 1990). Within the U.S. economy, fossil fuels consumed to generate electricity are the largest source of carbon dioxide, emitting about 35 percent of the total, followed by fossil-fuel use for transportation, industry, and residential and commercial buildings. Electricity's share is so large in part because about 60 percent of electricity is generated from coal. For each unit of energy obtained in combustion, coal emits some 40 percent more carbon dioxide than oil and almost 100 percent more carbon dioxide than natural gas.

In recent decades, several countries, including the United States, have taken steps to limit emissions of various types of air pollution caused by fossil fuels. Lead was phased out from gasoline in the seventies and eighties, automobiles were required to be equipped with catalytic converters, and power plants were subject to increasingly stringent controls on emissions. More recently, in 1990, amendments to the U.S. Clean Air Act required major additional reductions in sulfur and other emissions from various sources, to be phased in over the next several years. And most encouraging of all, the world community has reached agreement to phase out production of CFCs (although because of the long atmospheric lifetime of these chemicals, stratospheric ozone will continue to be depleted for decades to come).

For all of this, however, few concrete steps have been taken to limit carbon dioxide and other greenhouse gases (aside from CFCs). World fossil-fuel consumption and carbon-dioxide emissions have almost quadrupled since 1950, and without major changes in energy policies they are likely to continue expanding, possibly as much as doubling by 2025 (EPA 1990). Fossil-fuel consumption in the United States is predicted to increase 15 to 25 percent by 2010, if present trends continue (EIA 1991a). Even without the threat of global warming, it will be difficult to sustain these trends without causing excessive damage to the environment. As fossil-fuel use grows, it will become increasingly costly for communities to meet clean-air goals through conventional pollution-control strategies, necessary though they will be. Moreover, such strategies will do little or nothing to reduce emissions of greenhouse gases. A new approach to protecting the global environment is necessary, one that goes to the root of the problem: our society's addiction to fossil fuels.

Energy Choices

The need to reduce the risk of greenhouse warming presents all nations with a momentous challenge. Just to stabilize atmospheric concentrations of carbon dioxide at current levels will require reducing world carbon-dioxide emissions by at least 60 percent (IPCC 1990). In fact, industrial countries will have to reduce their emissions by an even larger fraction to allow less-developed countries room to industrialize and modernize their societies. Such a profound restructuring of the world energy economy cannot be accomplished on a time scale shorter than decades. Nevertheless, action should begin immediately, as every decade of delay will commit the earth to additional warming and make the eventual, inevitable adjustment in world energy consumption patterns more abrupt and wrenching.

Any proposal to move U.S. energy policy in this new direction is bound to provoke controversy, and indeed a raging debate is already under way. Some argue that the scientific uncertainties surrounding global warming are so large as to preclude major policy changes at this stage; further research is all that is called for. This view prevails within the U.S. government, which has so far resisted setting goals for greenhouse-gas reductions, as several other countries have done. But although it is possible that global warming will be less severe than climate models now predict, it is equally possible that the warming will be greater, and the consequences of *that* mistake could be disastrous. Moreover, any reduction in U.S. fossil-fuel use would have other important benefits, such as reducing air pollution and limiting oil imports and energy expenditures.

These considerations make a strong case for, at the very least, a "no regrets" strategy to reduce fossil-fuel use, one that adopts measures that make sense from an economic and environmental perspective regardless of whether global warming is accepted as fact. As we will see in the next chapter, such a strategy could indeed accomplish a great deal while at the same time reducing the danger of greenhouse warming.

Four principal strategies are available for reducing fossil-fuel use and carbon-dioxide emissions: switching from coal and oil to natural gas; improving energy efficiency; expanding the use of nuclear power; and developing renewable energy sources. The first of these should be encouraged because it could yield immediate benefits, including reduced pollution (since natural gas burns more cleanly than coal or oil). World reserves of natural gas are limited, however, so it cannot be a lasting solution. Of the other three strategies, energy efficiency and nuclear power have received the most attention in recent years, while renewable energy sources have been much less in the public eye. To understand the role renewables could play, we need to take a look at the other options first.

Energy Efficiency

The past 18 years have seen a revolution in the efficiency of energy use in the United States. For two decades before oil and other energy prices began rising in the early seventies, energy consumption had marched virtually in lockstep with economic growth. But from 1973 to 1986, this pattern changed as consumption stayed almost constant while the economy expanded almost 40 percent. Much of this change was due to

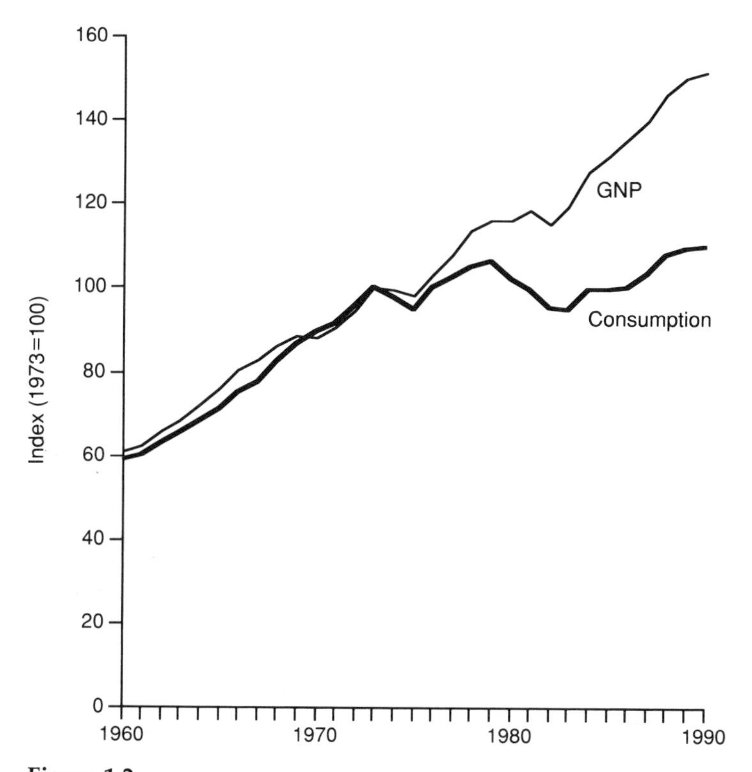

Figure 1.2
After the oil shock of 1973, U.S. energy consumption stayed relatively constant even as economic output (gross national product, or GNP) grew nearly 40 percent. Source: EIA (1991b).

improvements in the efficiency of products ranging from automobiles to air conditioners; the rest was the result of structural shifts in the economy (e.g., a move to less energy-intensive industry), behavioral factors (e.g., turning down thermostats), and a switch to fuels like natural gas that are produced and used more efficiently than coal and oil. Without these changes, it is estimated that the United States would be spending about $150 billion more for energy annually than it does today and would be importing nearly twice as much oil (Schipper, Howarth, and Geller 1990, Chandler et al. 1988).

Rising fuel prices had much to do with the increased efficiency, but government actions helped, too. The creation of Corporate Average Fuel Economy (CAFE) standards was probably the main driving force that led to a doubling of the efficiency of new cars sold in the United States, from 13 miles per gallon (mpg) in 1973 to around 28 mpg today. Likewise, appliance efficiency standards are at least partly responsible for a doubling of the efficiency of new refrigerators.

Despite this progress, vast quantities of energy are still being wasted. The United States uses about twice as much energy per dollar of gross national product as Japan, West Germany, and some other advanced industrial nations. Geographic and climate differences amongst these countries account for part, but only part, of this difference in energy use. For example, European countries and Japan are more densely populated than the United States, thus implying lower energy consumption for transportation, but they have also put more investment into efficient urban mass transit and intercity rail systems. The principal factor accounting for such differences is the price of energy, which is generally much higher in other countries than in the United States. For example, gasoline is taxed at a rate of several dollars per gallon in most European countries, thus creating a strong incentive for Europeans to buy more efficient cars, drive less, and use more mass transit.

Many possibilities exist for improving energy efficiency, ranging from advanced manufacturing processes to automobiles getting 60 mpg or more. Just as important as identifying these potential energy savings, however, is determining how much it would cost to capture them. Estimates of potential *cost effective* savings vary widely according to the method and assumptions used. On the one hand, most conventional "top-down" energy-economic analyses, which apply macroeconomic principles to aggregate measures of economic performance such as energy price, supply, and demand, indicate that achieving any significant energy savings would be very costly. Some studies have suggested, for example, that reducing fossil-fuel use by 20 percent would cost the U.S. economy on the order of $200 billion per year, an increase of 50 percent over the current national energy bill (Nordhaus 1989, Manne and Richels 1989). On the other hand, the less traditional "bottom-up" approach, which examines in detail the array of technology options available for every end use (space heating, transportation, and so on), tends to reach precisely the opposite conclusion. As two leading practitioners of this sort of analysis put it, "the econometricians have the amount [of monetary costs] about right but the sign wrong; using modern energy-efficient techniques . . . would not cost but save the United States on the order of $200 billion a year" (Lovins and Lovins 1991).

One critical drawback of the top-down approach is that it is based on historical relationships between price, supply, and demand, and is consequently unable to reflect changes in the variety of cost effective technologies available or to investigate the impacts of policies (like efficiency standards) that do not directly affect the price of energy. In addition, the approach ignores details about specific end uses of energy, which can have a major impact on overall energy use (Williams 1990). The bottom-up approach, on the other hand, ignores relationships between important macroeconomic parameters and thus tends to present an unrealistic picture of how the economy would respond to changing market conditions. For example, reducing fuel use would tend to reduce fuel prices, resulting in greater monetary savings but also greater demand for energy services; these effects are usually not considered in bottom-up models.

Thus, reality may lie somewhere in between the two extreme examples cited above. Yet the fact that West Germany and Japan do so well economically while using half as much energy per dollar of output as the United States suggests that reducing energy use need not place a heavy burden on the U.S. economy. Some policies that would lead in this direction include further raising CAFE standards on new motor vehicles, increasing federal and state gasoline taxes, reforming electric-utility regulations to foster investment in end-use efficiency, and placing more stringent efficiency standards on electric appliances, lighting, and other products. According to one estimate (firmly in the bottom-up camp), by the year 2000 policies such as these could result in a 7 percent reduction in fuel use below 1988 levels (18 percent below business-as-usual projections) with a net annual savings of $75 billion (Geller 1989).

Even greater efficiency gains may be possible over the long term. One study indicates that with current technologies industrial countries could cut their per-capita energy use by one-half and total energy use by one-third over 30 years. Over the same period, less-developed countries could raise their standard of living to the level of 1970s Western Europe by adopting more efficient technologies for cooking, lighting, and other basic needs. With these changes, world energy consumption would be only slightly higher in 2020 than it is today or about half of what it is projected to be otherwise (Goldemberg et al. 1987).

But, as even this optimistic study indicates, energy efficiency alone will not be enough to reduce world fossil-fuel use substantially below today's level because of the offsetting effects of population and economic growth. To achieve the goal of stabilizing global climate, it will be necessary to replace at least 50 percent of current fossil-fuel use with alternative energy sources that do not emit greenhouse gases. One option for doing so is to increase the use of nuclear power.

Nuclear Power

During the seventies, nuclear power was heralded as a promising means of reducing U.S. dependence on imported oil. Indeed, in the past 10 years the production of electricity from nuclear power plants has multiplied sixfold, and as of 1990 there were 111 plants with operating licenses supplying almost 20 percent of the electricity generated in the United States (EIA 1991b).

Despite this record of expansion, the future of nuclear power in the United States is in serious question. No new plants have been ordered since 1978, and all those ordered since 1973 have been canceled. Once the few plants under construction are completed, no new ones are likely to go into service for at least 10 years, as it takes that long to plan, construct, and obtain an operating license for a new plant. For this reason alone, nuclear power can make no immediate contribution to reducing fossil-fuel use, at least in the United States.

The two key obstacles confronting the nuclear industry concern safety and cost. Ever since the Three Mile Island accident in March 1979 — in which, for the first time, a reactor core partially melted, nearly causing a catastrophic release of radioactivity into the environment — the perception by many people that nuclear plants are unsafe has inhibited plans to build additional plants. According to a Louis Harris poll conducted in late 1988, 61 percent of Americans are opposed to the construction of

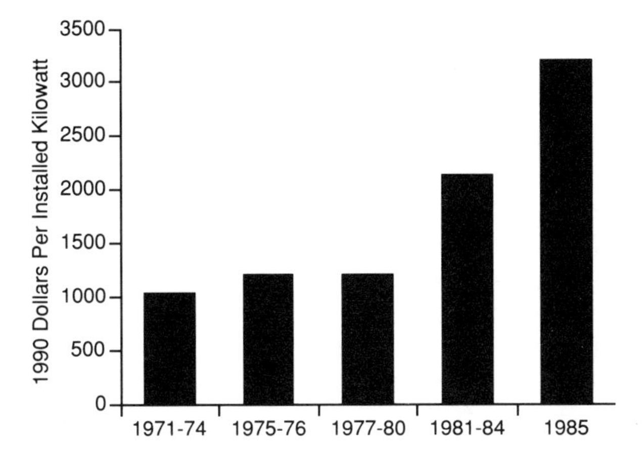

Figure 1.3
The average cost of nuclear power plants rose sharply in the 1980s because of new safety requirements, more complex power plant designs, and construction and licensing delays. Converted to 1990 dollars from EPA (1990).

more nuclear plants, while 30 percent are in favor — almost exactly the reverse of opinions found in answer to the same question 10 years earlier (Harris Poll 1989). At the same time, electric utilities and private investors have become disenchanted with the rapidly rising costs of nuclear power, which was once predicted to become "too cheap to meter." Nuclear plants completed in recent years have cost over $2,500 per installed kilowatt of capacity, over twice the cost of a typical coal-fired power plant.

The two issues, cost and safety, are not unrelated. Changing safety regulations requiring expensive design modifications and backfits have added to the cost of new reactors. The nuclear industry, however, was experiencing problems of low performance and escalating costs well before the Three Mile Island accident (Komanoff 1981). At least part of the industry's trouble appears to be caused by ineffective management and regulation, for U.S. reactors have historically performed poorly compared to West European and Japanese reactors of similar or identical design (Hansen et al. 1989). The situation was not helped by the U.S. government's decision in the fifties to adopt the complex and inherently unsafe light-water reactor (LWR) technology (developed for the Navy's nuclear submarines) for the civilian nuclear power program (Kendall 1991).

With construction of new light-water reactors at a virtual standstill, attention has been focused on the possible development of safer and more economical reactor designs to replace them. Some companies are concentrating on simplifying and standardizing LWR designs to improve their performance and safety. Westinghouse and General Electric are each developing an advanced LWR that runs at lower temperatures, has a capacity of 600 megawatts (MW) as opposed to the 1,000 MW or more typical of today's reactors, and relies on an enormous tank of water on top of the reactor to supply emergency cooling in the event of a loss-of-coolant accident. The companies expect these reactors to be certified for licensing sometime in the mid-1990s, but whether any utilities will buy them remains to be seen (WRI 1990).

Three quite different types of reactors are also under development in the United States and abroad: the Process Inherent Ultimate Safety (PIUS) reactor, of Swedish invention; the modular gas-cooled reactor (MGR), developed by American and West German firms; and the Integral Fast Reactor (IFR), based on a design by the Argonne National Laboratory. All of these designs depend to some degree on safety features requiring comparatively little human or mechanical intervention to prevent, or at least delay, core meltdown and the release of radioactivity. They are also relatively small reactors, with planned capacities of 100 MW to 450 MW. Smaller reactors have the advantages that they would require less capital investment and financial risk than larger reactors and their construction could be tailored more closely to electricity demand. None of these designs has yet been demonstrated in a commercial reactor, however, and it will be several years before their true cost and safety can be determined (MHB 1990).

Whether or not advanced reactor designs prove successful, other important problems call into question the feasibility of greatly expanding reliance on nuclear power. All nuclear power plants produce hazardous, radioactive wastes, which must be safely stored for up to hundreds or thousands of years. At present, most plants store their wastes on-site, but this is only a temporary solution. The United States has initiated investigations for a permanent waste repository to be constructed underground in Yucca Mountain, Nevada. The repository, which would hold wastes generated by existing commercial plants, is scheduled to be opened in 2010 at the earliest.

The Yucca Mountain project has already been delayed several years, however, and is facing growing opposition from Nevada citizens. The much smaller Waste Isolation Pilot Project (WIPP) in New Mexico, which is intended to test the concept of deep-underground storage using wastes from nuclear-weapons production facilities, has also been delayed, and questions concerning the danger of water intrusion into salt caverns, which could result in the eventual contamination of local groundwater, have not been resolved. Given the engineering difficulties and mounting political problems associated with siting waste facilities, it is unknown whether any new sites can be found to store wastes generated by plants constructed in the future.

An additional risk of expanding reliance on nuclear power is the possibility that nuclear materials will be diverted by countries or terrorist organizations for the purpose of making nuclear weapons. The risk of such diversions from the United States is currently very low because the "once-through" uranium fuel cycle of commercial reactors does not generate materials directly usable in weapons. Because world uranium reserves are limited, however, a global energy economy relying heavily on nuclear power would require the reprocessing of spent fuel and the construction of breeder reactors to produce plutonium. In contrast to the low-enriched uranium used in existing commercial reactors, plutonium can be readily used to make nuclear weapons. In a nuclear-based economy, large quantities of plutonium fuel would have to be transported on roads and railways and across oceans, posing a potentially serious risk of material diversion (not to mention accidents). It was in part because of this danger that the United States decided to forgo fuel reprocessing in the 1970s (Lipschutz 1980).

In sum, for at least the next decade, nuclear power plants cannot be counted on to supply much more energy in the United States than they do today. Indeed, their contribution may begin to decline after the turn of the century as aging plants are retired, if new plants are not ordered in the near future. Nuclear power may fare somewhat better in other countries, but probably not substantially. The 1986 Chernobyl disaster in the Soviet Union sparked a reevaluation of the role of nuclear power in the energy plans of many countries, and this is reflected in a marked worldwide slowdown of plant construction. While a greater role for nuclear power cannot be ruled out in the long term, this will require the resolution of serious problems concerning the safe disposal of radioactive wastes and the prevention of nuclear-weapons proliferation.

Note

1. CFCs play a dual role in climate change and stratospheric ozone depletion. The gases absorb infrared radiation from the earth's surface with very high efficiency, thus contributing to global warming. When they reach the upper atmosphere, the CFC molecules are broken down by sunlight, and the individual chlorine atoms react catalytically with other compounds to destroy ozone. Ozone is itself a greenhouse gas, and thus these two effects tend to counteract one another.

2 The Renewable Alternative

Renewable energy sources are often regarded as new or exotic, but in fact they are neither. Until quite recently in human history, the world drew most of its energy from the sun, either directly from sunlight or indirectly through the natural processes that generate winds, rivers, and plants. In the early 19th century, the most common source of energy in the United States was firewood. In areas where streams were plentiful, water power was used to thresh grain and mill lumber, and on farms, windmills for pumping water were a common sight. Houses were often oriented to capture sunlight in winter and provide shade in summer, and storage tanks were painted black so that water would be heated by sunlight. Most of these practices were gradually abandoned as fossil fuels — first coal, then oil and natural gas — came into wide use.

Renewable energy was not abandoned completely. The power of rivers was harnessed in the early part of this century to generate electricity, and today hydroelectric facilities supply 10 to 12 percent of the electricity generated in the United States (about half the contribution of nuclear power). In addition, some 5 million households burn wood as their primary source of heat, while the pulp and paper industry meets over half of its energy needs by burning process wastes. In all, about 8 percent of U.S. primary energy supply (that is, including energy consumed in the production of electricity) is provided by renewable sources.

Interest in other renewable energy technologies was revived in the seventies because of rising energy prices and growing concern over U.S. oil security and pollution caused by fossil fuels. As one barometer of public interest, federal funding for renewable energy research and development (R&D) soared from $75.1 million in fiscal 1975 to $718.5 million in fiscal 1980.[1] In the process, many new technologies were developed and old ones improved, and some even attained a measure of commercial success. Wind turbines, for example, were installed in large numbers in California in the early eighties, and solar collector sales jumped from virtually nothing to several hundred thousand units a year. As the sense of energy crisis faded, however, so did interest in developing alternatives to fossil fuels. R&D funding steadily declined, reaching a low of $114.7 million in fiscal 1989 — a cut of almost 90 percent below the 1980 level, if inflation is taken into account. Although in the past three years this trend has reversed, the fiscal 1992 R&D funding of $203.7 million is still far below its peak of a decade earlier (Sissine 1992). Solar collector sales are down 60 percent from their 1984 peak, and very few new wind turbines are expected to be installed in 1992.

Yet the advantages of renewable energy sources — particularly wind, solar, biomass, and geothermal — are, if anything, more compelling today than ever before. The technologies that have been developed, ranging from wind turbines and photovoltaic cells to liquid fuels derived from biomass to new architectural approaches and components for the use of solar energy in buildings, are of startling versatility. Most produce few or no pollutants and hazardous wastes. Drawing entirely on domestic resources, they are immune to foreign disruptions such as the 1973 Arab oil embargo, and they provide a hedge against inflation caused by the depletion of fossil-fuel reserves. Their development would very likely result in a net increase in employment, as renewable energy industries generally require more labor, per unit of energy produced, than coal, oil, and natural-gas industries (Flavin and Lenssen 1990, Hall, Cleveland, and Kaufmann 1986).

Most important, resources of renewable energy are vast and, if properly managed, virtually inexhaustible. Sunlight falling on the U.S. landmass delivers about 500 times as much energy as the United States consumes. The annual wind energy potential of just three states, North Dakota, Montana, and Wyoming, is equal to all of the electricity used in the United States in 1990. Biomass and hydropower resources, though smaller, are also substantial. The geothermal resource — including as-yet undeveloped resources such as hot dry rock — is estimated to be several hundred times greater than proven U.S. reserves of coal.[2]

In practice, of course, only a fraction of these resources could be exploited because of constraints on available land, the efficiencies of energy conversion, environmental considerations, and other factors. Even considering these limitations, however, the amount of energy that could physically be recovered from renewable sources almost certainly far exceeds current and foreseeable U.S. energy demand. Solar and geothermal energy offer perhaps the greatest long-term potential. Solar collectors covering less than 1 percent of U.S. territory — one-tenth the area devoted to agriculture — could make more energy available than the United States consumes in a year. Hydropower has the least room for expansion, since most of the attractive sites for hydropower have already been developed, and many of the rest are barred from development by environmental regulation.

A Decade of Progress

Despite the impressive potential of renewable energy sources, they have been virtually ignored by most mainstream energy analysts, many of whom regard them as expensive and impractical. This opinion was at least partially accurate in the seventies, when the modern renewable energy industry was just getting started. Wind turbines suffered mechanical failures, solar collectors were costly, and geothermal development was still in its infancy. At the same time, the simplest of all solar technologies — "passive" heating of homes — was a technical and economic success, but largely ignored.

Yet the costs of renewable energy technologies have declined dramatically since then, and their reliability has been proven in demonstrations and commercial operation. This progress — all the more remarkable

Table 2.1
Current utilization of renewable energy sources and recoverable potential, in exajoules (EJ) of primary fuels displaced per year. For comparison, the United States consumed 89 EJ (84 quadrillion Btu, or quads) of primary energy in 1991. For sources, see technology chapters.

Resource	Current Use	Recoverable Potential
Wind (High Quality)	0.04	6.1
Wind (Medium Quality)	—	107.2
Biomass	3.45	35-71
Hydro	3.31	4.0
Solar	0.08	>100
Geothermal	0.18	>100

considering the lack of official support for renewable energy in the eighties — calls for a reassessment of renewable energy's potential role in meeting U.S. energy needs.

Wind turbines are a good example of the growing competitiveness of renewable energy technologies. The cost of electricity generated at typical sites in California, where most wind development has taken place, has declined from over 25¢/kWh in 1981 to around 5-8¢/kWh today, depending on wind speeds and other conditions, and when the next generation of wind machines becomes available in 1993, the price could fall to as low as 4-6¢/kWh.[3] (For comparison, new coal-fired power plants generate electricity at a cost of around 5¢/kWh.) Reliability problems affecting early wind-turbine designs have been largely resolved, and mature and well-maintained systems are available 95 to 98 percent of the time.

Geothermal energy is now well established as a source of electricity in the United States, although it is currently limited to a few western states where high-temperature hydrothermal reservoirs containing steam or hot water are found. The earth's heat could eventually be used to generate electricity in other parts of the country if it proves commercially feasible to tap into geopressured fluids and hot dry rock, two approaches that have already been demonstrated on a pilot scale. Biomass could also become a major source of electricity, especially if new, more efficient combustion technologies (some of them similar to those developed for the Department of Energy's "clean coal" program) are developed and demonstrated.

Other renewable sources of electricity, such as solar-thermal power plants and photovoltaic cells, have seen cost decreases of 60 to 90 percent in the past decade. Though still too expensive (at least by conventional measures) to make a major impact in the electricity market, their costs are expected to drop further with additional improvements in technology and economies of scale. Small wind turbines and photovoltaic systems are already the technologies of choice for many remote applications, such as supplying village power in developing countries or powering navigation aids and communication relays in the United States. Ocean energy systems are somewhat more speculative, but studies suggest they could find important niche markets in suitable locations, such as tropical islands.

For applications requiring direct heat — over half of the end-use energy consumed in the United States — solar and geothermal energy are viable options today in some locations and could be much more widely developed in the future. Passive solar building design, which uses a building's structure to capture and store solar energy, is an elegantly simple approach that can substantially reduce energy use in new buildings at little or no extra cost. Solar concentrating systems now on the market can supply heat for commercial uses at an annualized cost (in sunny climates) only slightly higher than the current price of natural gas, and, like residential solar water heating systems, their cost could drop by one-third or more simply through the production of larger numbers of units. Where available, low-cost geothermal energy is already being used to provide heat for buildings and industry, and this practice could spread if new approaches — particularly hot dry rock — are developed.

Developing renewable substitutes for gasoline and other transportation fuels is perhaps the most difficult challenge, but even here there is promise of a solution. Ethanol can now be produced from wood and other plants at about twice the refinery cost of conventional gasoline. With continued improvements in conversion processes and methods of cultivating biomass feedstocks — and with projected increases in the price of gasoline — the two fuels could become roughly competitive around the turn of the century. Forestry and agricultural wastes, as well as plants and trees grown specifically for energy, would supply the raw materials. Further in the future, cars powered by renewable hydrogen or electricity (provided by low-cost renewable sources, such as biomass or photovoltaic cells) are a realistic possibility.

What happens when the sun goes down or the wind stops? Conventional wisdom holds that energy storage will be needed to keep solar and wind power flowing reliably, thus greatly increasing the costs of these energy sources. The basic fallacy in this complaint, however, is that it treats intermittent renewable energy sources in isolation from all others, thus ignoring the important advantages of integrating them into larger and more diverse energy systems.

In some applications, for example, sufficient storage or backup capacity already exists to ensure reliable power from intermittent renewable sources. Electric utilities maintain a reserve capacity (typically 20 percent in excess of peak demand) to allow for unexpected plant shutdowns. This reserve should suffice until solar and wind energy constitute at least a few percent, and possibly as much as 20 percent, of the total electricity supply — a level of market penetration that will not be achieved for decades. Furthermore, hybrid energy systems drawing on both renewable and fossil sources can provide reliable power while greatly reducing fossil-fuel consumption. There are power plants operating in Southern California, for example, that run on 75 percent solar energy and 25 percent natural gas and supply reliable power year-round. Natural gas can also supplement solar energy in residential, commercial, and industrial heating applications for little extra cost. Finally, in those cases where storage is needed, numerous options exist, some of which, such as compressed air energy storage, are likely to be practical and relatively inexpensive in many parts of the country (see chapter 8).

This is not to say that there are no significant technical challenges to be overcome. To be sure, photovoltaic systems need to become more efficient and less expensive to manufacture, the yields of energy crops need to be improved, and hot dry rock technology needs to be demonstrated on a commercial scale. But in most cases, as we will see in the technology chapters, these and other challenges appear surmountable, given sufficient government and industry investment. At heart, the major obstacles standing in the way of an orderly transition to a renewable energy economy are not technical in nature, but concern the laws, regulations, incentives, public attitudes, and other factors that make up the energy market.

The Market for Renewable Energy

For all of the progress made by renewable energy technologies in the past decade, many are still struggling for commercial success or remain simply stuck in the laboratory. A glut of low-cost fossil fuels, sharp cutbacks in federal support, and above all an energy market obsessed with the short term and insensitive to the environmental and social costs of conventional fuels, have combined to slow industry growth. If present circumstances remain unchanged, according to official forecasts, renewable energy sources will provide only about 13 percent of U.S. energy supply in 2010, only slightly more than today's 8 percent contribution (EIA 1991a).

Market Barriers

One basic difficulty is that the public remains generally ignorant of the nature and value of renewable energy sources. This problem has been exacerbated by horror stories of, for example, wind turbines losing their blades and solar homes overheating, which have raised doubts about the practicality of these energy sources. Although most such problems have been solved or could be avoided in the future through proper regulation, it remains difficult for companies marketing renewable energy technologies to win the recognition and confidence of customers and investors.

In addition, homeowners know little about passive solar heating or solar water heating. Electric utility managers are unsure of how well renewable energy sources — particularly intermittent wind and solar power — would work on their utility systems. Industry managers are even less inclined to invest in new sources of energy with which they have no experience and which do not lead directly to improvements in the products they sell. Even when people are interested in renewable energy, they often find it difficult to obtain reliable information about available technologies and resources.

Exacerbating these problems is the fact that most renewable energy technologies cost a lot up front while providing savings down the road in the form of lower fuel costs. This creates a significant capital cost barrier that further discourages interest in renewable sources. A solar water heater, for example, may cost $2,500 to purchase and install, whereas a conventional water heater may cost no more than $800. Similarly, a solar power plant may cost $2,500 to $3,000 per kilowatt of capacity, whereas a conventional power plant costs anywhere from $400 to $1,200 per kilowatt. The difference is that the solar technologies cost very little to operate, whereas the major cost associated with conventional technologies is usually fuel, which will be paid later.

Even our tax system tends to penalize capital-intensive renewable energy investments. Businesses must pay sales taxes on the purchase of capital equipment as well as taxes on profits to investors, whereas operating expenses (including fuel costs) can be wholly deducted. In effect, this makes a dollar invested in capital equipment more expensive than a dollar invested in operations.

Research and Development

Since renewable energy sources represent largely new technology, investment in research and development is of critical importance to their success. Yet as we have seen, federal funding for renewable energy R&D declined throughout the eighties, and despite substantial increases in fiscal 1991 and 1992, it still constitutes less than 10 percent of total federal funding for energy supply R&D.[4]

Even when new technologies have been developed to the point of being ready for commercial testing and deployment, they have not been picked up by industry. During the eighties, in fact, the Department of Energy eschewed programs designed to foster transfer of technology to industry on the grounds that such programs would interfere in the "free market" and usurp the role of industry. Instead, emphasis was put on basic research. Most renewable energy companies, however, do not have the financial strength or technical know-how to turn component technologies into reliable and marketable systems. Photovoltaic cells are one example: Even though scientists' understanding of the inner workings of photovoltaic cells has advanced dramatically in the 1980s, the same cannot be said for the industry's ability to produce reliable, low-cost photovoltaic modules (integral collections of cells) on a large scale. Other promising technologies, such as tapping the geothermal energy of hot dry rock or the thermal energy of oceans, are virtually doomed unless the federal government provides technical support and funding for commercial-scale demonstrations.

Environmental and Social Costs

Lastly, markets largely ignore the environmental and social costs of fossil fuels, which include such effects as pollution, greenhouse warming, and military expenditures to defend Persian Gulf oil supplies. This means, in effect, that consumers invest excessively in fossil fuels to the detriment of society as a whole. Finding some way to "internalize" the environmental and social costs of energy sources is essential to promoting greater renewable energy use. An additional tax of 50 cents per gallon on gasoline, for example, could accelerate by several years the introduction of biomass-derived fuels such as ethanol, while a premium of 1¢/kWh on coal-generated electricity would make wind power far more attractive. (See chapter 9.)

The environmental and social costs of fossil-fuel use (with the exception of direct subsidies and military expenditures) are difficult to determine with any precision, although many studies have attempted to do so (Chernich and Caverhill 1991, Ottinger et al. 1990). For example, a review study commissioned by the American Lung Association concluded that the health impacts of air pollution alone cost the economy anywhere

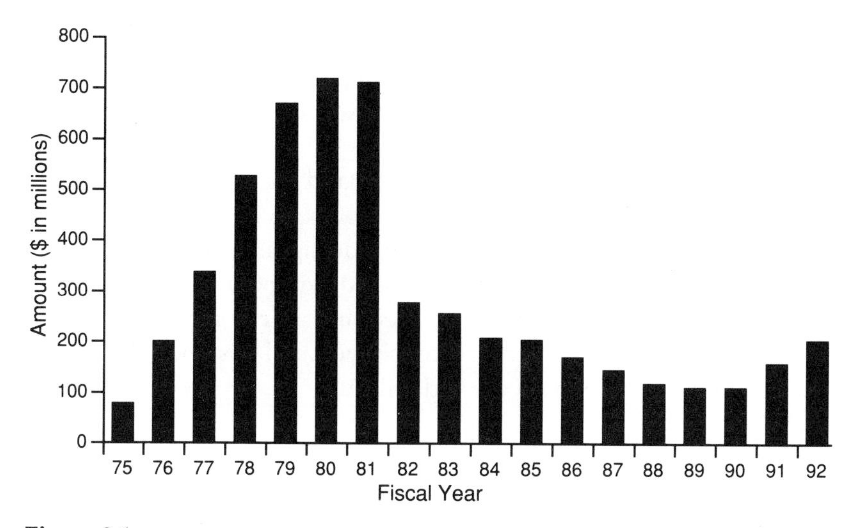

Figure 2.1
Research and development (R&D) funding for renewable energy has gone up and down with the political times, making it difficult to conduct a coherent program. Source: Sissine (1992).

from less than a billion dollars to more than $100 billion annually. One reason for the broad spread is the uncertainty surrounding estimates of health effects; another is wide disagreement over the economic value to be placed on human life (Cannon 1990).

Because of conceptual and practical difficulties such as these, the trend (at least in the area of electric-utility regulation) is increasingly toward relying on the cost of *controlling* pollution, rather than the cost of damage, on the assumption that the control cost more accurately reflects the value society places on reducing pollution. Depending on the emission source, the total pollution cost calculated in this manner can be quite high. For example, using emissions values adopted by Nevada for its utility planning process, pollution adds 2¢/kWh to the effective cost of electricity from a conventional coal-fired power plant, or 4.4¢/kWh if greenhouse-gas emissions are taken into account (Bernow et al. 1990).

Pollution is not the only environmental or social cost that should be considered, nor are fossil fuels the only culprit. Expenditures for military forces assigned to protect the Persian Gulf are estimated to add at least $2.50 to the effective cost of a barrel of oil (Broadman and Hogan 1988). The use of nuclear power exposes the public to the risk of a major accident, the potential cost of which (because of federal limitations on liability for nuclear accidents) is not fully reflected in the price charged for nuclear-generated electricity (Dubin and Rothwell 1990). As we will see, renewable energy sources can have significant environmental and social impacts, as well, though with some exceptions they are far smaller than the impacts of fossil fuels.

At the same time, some environmental and social costs are already accounted for implicitly in the market, if not explicitly in energy prices. Environmental regulations, for example, add significantly to the cost of energy from certain sources, in some cases driving them completely out of contention. To a considerable degree this is true for hydropower, whose development has been greatly slowed by laws and regulations intended to protect scenic rivers and wildlife.

Valuing the Future

The many barriers to renewable energy development all point to a central issue: What value should society place on future, as opposed to present, costs and benefits of energy use? Economists and accountants express the concept of time value of costs and benefits as a discount rate, which is defined as the percent decrease in the value of a dollar earned or spent in one year compared to the value of a dollar earned or spent the year before. A higher discount rate means that less value is placed on future savings or expenditures.

What discount rate is assumed has a considerable impact on the perceived attractiveness of renewable energy investments. A "real" (i.e., inflation-adjusted) discount rate of 8 percent, for example, means that a dollar earned 10 years from now has a present value of just 43 cents. A real discount rate of 3 percent, in contrast, means that a dollar earned 10 years from now has a present value of 74 cents. After 30 years, the gap grows even wider. The present value of a dollar earned in 30 years is only 8 cents if the discount rate is 8 percent, whereas it is 40 cents, or five times as large, if the discount rate is 3 percent. Thus, a solar water heater, which generates long-term earnings in the form of lower fuel costs, will be perceived as a much more attractive investment if it is evaluated using the lower discount rate rather than the higher discount rate.

Of course, the choice of discount rate is not arbitrary. For the private sector, it is usually based on the cost of borrowing or raising capital (although other factors, such as the riskiness of the investment, may also be taken into account). This cost varies widely depending on the circumstances, however, and government regulations and incentives can make an important difference. For privately owned electric utilities, the annual cost of servicing a typical 50:50 mix of equity and debt is about 6 percent, in real terms, or 11 percent in nominal terms (that is, taking into account the standard assumption of a 5 percent per year inflation rate) (EPRI 1989c).[5] On the other hand, independent power producers (IPPs) — companies selling electricity to utilities — may have a much different (and usually lower) discount rate depending on their debt and financing structure.[6] This implies that some power plant investments will appear attractive to utilities and not to IPPs, and vice versa.

In general, a homeowner or small company wishing to invest in, say, a solar water heater faces a much higher cost of capital than a utility or other large company *unless* long-term financing or leasing arrangements are available. The interest rate on credit card debt, for example, is 18 percent in nominal terms, or 13 percent in real terms. But if the homeowner has the option to include the cost of a solar water heater in his or her home mortgage, then the interest rate may be as low as 9 to 10 percent (nominal) or 4 to 5 percent (real), that is, actually lower than the cost of capital for utilities. Thus, such a simple change in the rules governing mortgage financing — a long-standing proposal of the solar industry — would result in a factor of two or three decrease in the amount the homeowner would have to pay annually for a solar water-heating system.[7]

These examples suggest the critical importance of policies designed to lower the cost of capital for private investments in renewable energy sources. Society itself (represented by its government) may choose to assume a relatively low discount rate for renewable energy investments to reflect its interest in protecting future generations from damage caused by today's energy decisions. There is, obviously, an ethical dimension involved in this choice, for one may ask what right society has to value the well-being of people not yet born any less than the well-being of people now living. Thus, although in practice governments often choose to assume a discount rate equal to the prevailing interest rate on government bonds — usually 3 or 4 percent in real terms — one can at least make an argument to justify a lower discount rate — even zero.

Risk is another factor that should be taken into account in selecting the proper discount rate for evaluating energy investments (Awerbuch 1991b). One risk associated with fossil sources is that fuel prices may increase unexpectedly in the future, resulting in financial losses for the investor or electric-utility customer; a principal benefit of renewable energy sources (other than biomass) is that they provide protection against this risk. Out of ignorance, perhaps, many homeowners and small businesses simply assume that fuel prices will stay constant when they compare different energy investments. They calculate the *simple payback period* for an investment, which is the time needed to recover the initial cost through fuel savings at *current* prices. This approach obviously works to the disadvantage of products like solar water heaters.

Most larger businesses and financial investors consider risk factors routinely in their business decisions. Stocks and bonds, for example, are valued according to their perceived level of risk: High-risk investments such as junk bonds are expected to achieve a high average rate of return, whereas low-risk investments such as Treasury bonds are expected to achieve much lower returns. A typical investment portfolio will contain a mix of financial instruments designed to attain the highest possible return appropriate for the desired level of risk. Oddly, though, electric utilities do not practice this sort of risk adjustment in deciding which type of power plant to invest in. On the contrary, utility regulations usually mandate the use of a single discount rate for evaluating all prospective power plant investments, regardless of their comparative riskiness.[8]

Of course, utilities consider the projected rate of increase in fuel prices when evaluating power plant investments, but they usually give little consideration to the impact on consumers should fuel prices deviate substantially from the projections. In essence, they assume the future is set in stone, whereas in reality, fuel prices — especially gas and oil prices — very rarely follow expectations. The appropriate response of utilities and regulatory commissions to fuel-price uncertainties should be to invest in fuel diversity, that is, in a mix of fossil-fuel and alternative sources.

Nor is fuel-price instability the only risk to be considered. New environmental regulations may impose unexpected costs on utilities and industries by requiring stringent new pollution controls and other measures. For example, residents of several midwestern states will soon be forced to pay much higher electricity rates because the 1990 Clean Air Act Amendments will require utilities to cut sulfur emissions from coal-burning power plants. This pattern could well be repeated if utilities and industry continue to invest in fossil-fuel technology, only to find limits placed on carbon-dioxide emissions later on.

Measured against these risks, solar, wind, and other renewable energy investments seem attractive indeed. For these technologies, it might well be appropriate to apply a lower discount rate than usual for evaluating their cost effectiveness. Put another way, it might be appropriate for government (state or federal) to provide financial incentives or change regulations in order to lower the cost of capital for private investments in renewable energy. According to an analysis of a hypothetical photovoltaic power plant investment, for example, the appropriate risk-adjusted discount rate should be around 3 or 4 percent (in real terms). Making such an assumption renders an expensive investment suddenly an attractive one (Awerbuch 1991a).

Needless to say, most analytical assessments of renewable energy ignore these subtleties and assume (implicitly or explicitly) a single, usually high, discount rate. This leads very often to rather pessimistic conclusions about its potential.

The Potential of Renewable Energy

Given the many barriers to renewable energy development, progress along the road to a sustainable energy system for the United States will be very slow, at least until some crisis — such as a sudden and sustained rise in oil prices — forces a radical restructuring of the energy market. By then, of course, it may be too late to prevent severe effects from global warming and other environmental problems. A far better course would be to remove or overcome these barriers through changes in energy policy. Some policy proposals are discussed in chapter 9.

Assuming that such policies will be put into place, the question remains: How much can renewable energy technologies be expected realistically to contribute to U.S. energy supply without placing undue burdens on the economy? As in the debate over energy efficiency, answers to this question vary widely depending on method and assumptions. The conventional wisdom holds that renewable energy sources can make only a limited contribution because of high cost, the dispersed nature of these resources, and the variability of wind and sunlight. Typical of this view is the following comment by an electric-utility analyst:

> Direct solar and other [renewable] energy systems are expected to have only a limited role within the next 50-60 years in the developed countries. The fundamental limitation is the low average solar availability in these countries and the resulting poor economics of solar systems compared to other energy technologies. . . . I can project no more than 50,000 to 100,000 MW of capacity [implying] a potential carbon substitution of 0.05-0.1 gigatons [per year]. (Spencer 1991)

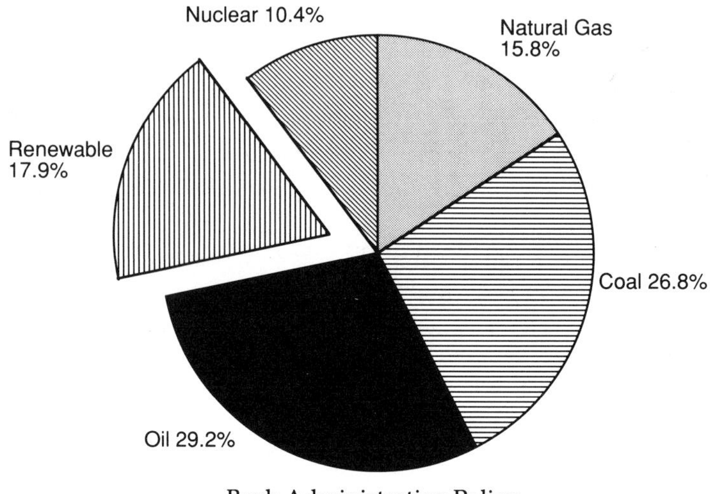

Bush Administration Policy
Total Consumption 129 Exajoules

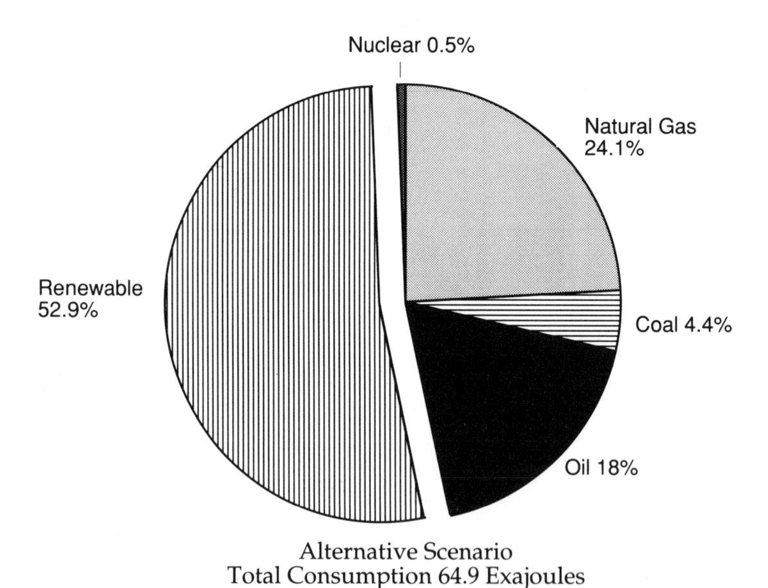

Alternative Scenario
Total Consumption 64.9 Exajoules

Figure 2.2
The Bush administration's National Energy Strategy, released in February 1991, envisions much higher energy consumption and greater reliance on fossil fuels and nuclear power in 2030 than an alternative strategy put forward by a coalition of public interest groups. Sources: NES (1991), UCS (1991).

In comparison, this analyst projects an increase of 100,000 to 1,000,000 MW in nuclear capacity over the same period.

Most such statements appear to be based on very little hard analysis and do not consider the possibility that market or policy conditions might change. The author quoted above may indeed be right that under conventional assumptions, and in the present market, renewable energy sources will not make a major contribution. But is it really in society's long-term interest to continue relying on fossil fuels indefinitely, and suffer the consequences of global warming, rising energy prices, and near-total dependency on Middle East oil? Clearly not — so a different approach is needed.

One such approach was presented in a study released in 1991 by a coalition of energy-advocacy groups — the Union of Concerned Scientists (UCS), Natural Resources Defense Council (NRDC), American Council for an Energy-Efficient Economy (ACEEE), and Alliance to Save Energy (ASE) — in conjunction with the Tellus Institute, an energy consulting firm based in Boston (UCS 1991). This study used the end-use oriented, bottom-up approach described earlier, but went one step further than most such studies by integrating both efficiency and renewable energy technologies into the analysis. The study constructed four energy scenarios for the United States covering the period 1988 to 2030: a Business-as-Usual scenario, designed to reflect current policies and trends and based on Department of Energy projections; a Market scenario, which selected energy technologies based on the goal of minimizing the cost of energy services purchased by consumers, assuming moderate market-penetration rates of new technology; an Environmental scenario, which assigned monetary values to the environmental impacts of energy use; and a Climate Stabilization scenario, which sought to meet predetermined targets for reduction of carbon-dioxide emissions into the atmosphere (a 25 percent reduction from 1988 levels by 2005, and a 50 percent reduction from 1988 levels by 2030).

Over a hundred energy-efficiency and renewable energy technologies were assessed. The principal conclusions for the most aggressive scenario, Climate Stabilization, were that:

- National energy requirements in 2030 could be cut nearly in half from the Business-as-Usual case, with renewable energy sources providing more than half of the energy supply.

- U.S. petroleum consumption could decrease to one-third of current levels.

- Carbon-dioxide emissions could be cut by more than 25 percent from 1988 levels by 2005, and by more than 70 percent by 2030.
- If these goals were met, consumers would receive a net monetary savings (fuel savings minus investment cost) of $2.3 trillion over the next 40 years.

A key assumption in this study was its adoption of a "societal" discount rate of 3 percent (real) for evaluating energy investments and calculating net savings. This had the effect of making investments with long-term payoffs more attractive and so boosted both efficiency and renewable energy technologies (especially the latter). When the net savings were recalculated assuming a "private" discount rate of 7 percent, they decreased to about $0.6 trillion.

Could energy efficiency and renewable energy sources play a major part in helping to slow and eventually stop global warming? Based on analysis like that described above, the answer seems to be yes. If all countries took full advantage of opportunities to improve energy efficiency, then global fossil-fuel use and carbon-dioxide emissions would grow slowly, if at all. And if, in addition, renewable energy sources were

Table 2.2
One possible scenario of U.S. energy supply in 2000 and 2030, in exajoules per year, assuming implementation of policies to encourage energy efficiency and renewable energy development. Source: UCS (1991).

		Year	
	1990	**2000**	**2030**
Biomass	3.45	7.65	18.38
Hydropower	3.31	3.61	3.77
Geothermal	0.18	0.57	3.79
Solar	0.08	0.60	4.59
Wind	0.04	0.68	3.63
Renewable Total	**7.06**	**13.10**	**34.52**
U.S. Consumption	89.0	82.7	65.3
Renewable Fraction	**7.9%**	**15.8%**	**52.9%**

developed to their full potential, fossil-fuel use and carbon-dioxide emissions could be cut well below today's levels, eventually approaching the 60 percent or greater reduction needed to stabilize carbon-dioxide concentrations at today's levels.

This was the conclusion of an Environmental Protection Agency study that examined a wide array of policy options for mitigating global warming (EPA 1990). Under the EPA's "rapidly changing world" scenario, for example, policies designed to promote energy efficiency and renewable energy sources were judged able to reduce the earth's warming commitment (temperature increase to which the earth will be committed) an estimated 27 percent by 2050 and 42 percent by 2100. Energy efficiency and renewable energy sources were about equally important in this analysis and together accounted for nearly two-thirds of the total warming reduction — 65 percent — deemed feasible in the next century. In contrast, nuclear power's potential contribution was judged to be less than one-third that of either energy efficiency or renewable energy sources alone. (Among other policies considered were changes in agricultural practices and reductions in the use of chlorofluorocarbons.)

One of the most important conclusions of the EPA study was that renewable energy sources could make up 30-45 percent of the projected global primary-energy supply by 2050. The study coordinated by the Union of Concerned Scientists suggests that they could be developed even more quickly in the United States because of the great technical and financial resources concentrated here. Starting almost immediately, the United States could begin following a transitional path to an energy economy relying almost entirely on renewable energy sources.

Of course, theoretical studies such as these will not settle the debate over the potential of renewable energy. One necessary step to arriving at a consensus will be to identify and resolve discrepancies between the "bottom-up" and "top-down" methods of analysis, which tend to reach sharply contrasting conclusions. Nevertheless, the studies are suggestive of the bright future for renewable energy *if* governments act to remove the market barriers hindering its advancement.

Notes

1. Research funding includes money for solar buildings, photovoltaics, solar thermal, biofuels, wind, ocean thermal, international programs, technology transfer, the National Renewable Energy Laboratory (formerly the Solar Energy Research Institute), geothermal, and small-scale hydropower. See Appendix B.

2. Because heat conducts through rock so slowly, geothermal energy is not strictly a renewable resource, though it is so large in absolute terms that it is usually counted as renewable. See chapter 7.

3. Except where otherwise noted, costs are cited in levelized 1990 dollars (i.e., discounted and averaged over the lifetime of the technology), and calculated using conventional electric-utility assumptions and methods. See EPRI (1989c).

4. Out of $2.533 billion spent for energy supply R&D by the Department of Energy in fiscal 1991, $157.8 million (6 percent) went to renewable energy, $391 million (15 percent) went to clean coal technology, $461 million (18 percent) to other fossil-fuel technology, $306 million (12 percent) to nuclear fission, $275 million (11 percent) to nuclear fusion, and the rest to various other programs (Sissine 1992).

5. In utility decision making, the weighted average cost of capital becomes the discount rate, which may then be used to calculate the present value of a power plant investment over its lifetime (including capital, fuel, and operating costs). Alternatively, the discount rate may be used to calculate the annualized, or levelized, cost per kilowatt-hour generated, which is the figure most often referred to in comparing different power plant investments.

6. Usually, independent power producers borrow a much larger share of their capital needs than utilities, resulting in a lower average cost of capital, since debt is less expensive than equity.

7. This example ignores the effect of depreciation, which will raise the effective cost of capital for the homeowner, as he or she will be forced to replace the system at the end of its practical lifetime (perhaps 20 years). Nevertheless, the basic principle holds true.

8. Utilities have become more sensitive to risk, however, as a result of low demand growth and many failed nuclear investments in the seventies and eighties. This is one important factor driving utilities away from conventional, large-scale coal and nuclear power plants toward what are perceived as less risky, small-scale, gas-fired plants. The latter raise significant fuel-price risks, however, which are currently largely ignored.

3 Solar Energy

In Aeschylus's play *Prometheus Bound*, the god of fire recounts how he first found the people of earth, unenlightened and lacking "the knowledge of houses turned to face toward the sun." Aeschylus was not referring to his own people, the ancient Greeks, who had a deep appreciation for solar energy — something that our civilization, for all of its technological achievements, lacks. Early Greek architecture incorporated what we now call passive solar design, orienting buildings to admit solar radiation and using thick masonry walls to capture the sun's heat and shade trees to keep buildings cool. In a more daring vein, the inventor Archimedes is said to have used a "burning mirror" to set fire to Roman galleys attacking the harbor at Syracuse while they were still beyond the range of bow and arrow, a story scientists still debate.

Contrary to popular belief, solar power did not spring ready-made from the oil crises of the 1970s. Its rich history includes Joseph Priestley's use of a concentrating lens to heat mercuric oxide and discover elemental oxygen in 1744. His contemporary, Antoine Lavoisier, built a solar furnace that achieved temperatures of 1,750°C. Another Frenchman, Augustin Mouchot, devised several solar-powered steam engines in the late 1800s, using silver-plated reflectors that could be turned to follow the sun, and connecting the solar receiver to a steam boiler. During the early part of this century, solar water heaters were quite popular in the United States, especially in southern climates where winter freezing was not a serious problem. But when cheap oil and gas became available in the 1920s and 1930s, sales of solar collectors declined, and by the middle of the century the industry had virtually disappeared (Cheremisinoff and Regino 1978).

Although soaring energy prices and shortages in the 1970s sparked renewed interest in solar alternatives, this revival proved short-lived. State and federal tax credits, designed in part to help reduce the tax preference afforded conventional energy sources, had helped boost the solar industry in the late seventies and early eighties, but by the late eighties most of these had been reduced or eliminated, and federal research and development funding was cut to the bone. When combined with declining oil and gas prices, these events drove many solar companies out of business.

Today the pendulum appears to be swinging back again, but slowly. This time the main impetus is a combination of growing public concern about the environment and a more mature industry striving to make solar energy economically competitive. Solar technologies have vastly improved in the past decade, with advances in every field — in passive solar design and construction techniques, solar cells, industrial heating, and home hot water systems, to name a few. But while poised to supply a much larger share of U.S. energy needs, the industry still faces major barriers, including continuing low energy prices, tax codes that are biased against capital-intensive technologies, and the failure of markets to account for the environmental impacts of energy use. Unless these barriers are removed, solar energy will very likely remain at the margins of the U.S. energy picture for at least another decade and quite possibly longer.

The Solar Resource

Ultimately, virtually all the energy we use, fossil or renewable, comes from the sun.[1] Fossil fuels were created hundreds of millions of years ago when vast forests died and partially decayed, with the organic materials slowly being compressed as sediments were deposited above them. Although these fossil-fuel reserves are immense by human standards, they represent only a minute fraction of the sunlight that struck the earth during their formation. An impressive statistic: The earth receives as much energy from sunlight in 20 days as is believed to be stored in the earth's entire reserves of coal, oil, and natural gas.[2]

It should therefore come as no surprise that solar energy is practically a limitless resource. Over 40,000 exajoules (EJ) of sunlight fall on the U.S. landmass each year, an amount equivalent to 500 times current U.S. energy consumption. Of course, because of the effects of latitude, weather, and local topography, different parts of the country receive different amounts of sunlight, but the variations are not as great as one might expect. From the cloudy Northeast to the sunny Southwest there is a variation of only about a factor of two in the average annual amount of sunlight received. Such differences ultimately translate into cost. In areas with more sunlight, less collector area is required to produce a given amount of energy, and for many solar systems the collector is the most expensive component.

Sunlight hitting a collector may arrive from many different directions. *Direct* sunlight travels straight from the sun to the collector with only slight diffraction and scattering of the rays in the atmosphere. *Diffuse* sunlight is scattered many times by clouds or haze. The distinction is an important one, since only direct radiation can be used by systems that concentrate sunlight with mirrors or lenses. Flat-plate collectors, on the other hand, can use both direct and diffuse radiation, as can passive solar design techniques for heating and lighting buildings. Naturally, some parts of the country are more cloudy and hazy than others: Albuquerque receives close to 70 percent of its sunlight as direct radiation, whereas in Boston the fraction is less than 50 percent.

Although the solar resource is enormous, it is much more widely dispersed than fossil-fuel resources. This is often taken to mean that a great deal of land will be needed for solar energy to make a significant contribution, but the issue is more complicated than that. True, some types of solar systems — in particular, multimegawatt central power plants — will have to be situated on large tracts of land (approximately one square kilometer for every 20-60 MW generated, depending on the conversion efficiency).[3] In reality, as we will see later in this chapter, large central power generation is only one of many possible ways to produce energy from sunlight, and perhaps one of the least promising. Small-scale, dispersed applications are a better match to the resource. These can take advantage of unused space on the roofs of homes and buildings and in urban and industrial lots. And in passive solar building design, the structure itself acts as the collector, so there is no need for any additional space at all. In short, the main constraints on solar use will be economic and technical in nature, not a shortage of land.

Solar Buildings

Of the three main types of solar systems — solar buildings, solar-thermal concentrating systems, and photovoltaic cells — solar buildings have received the least public attention and government funding. Yet they are vitally important because of the large amount of energy that goes to heating, cooling, and lighting buildings. Residential homes and commercial buildings account for approximately one-third of U.S. energy use, if

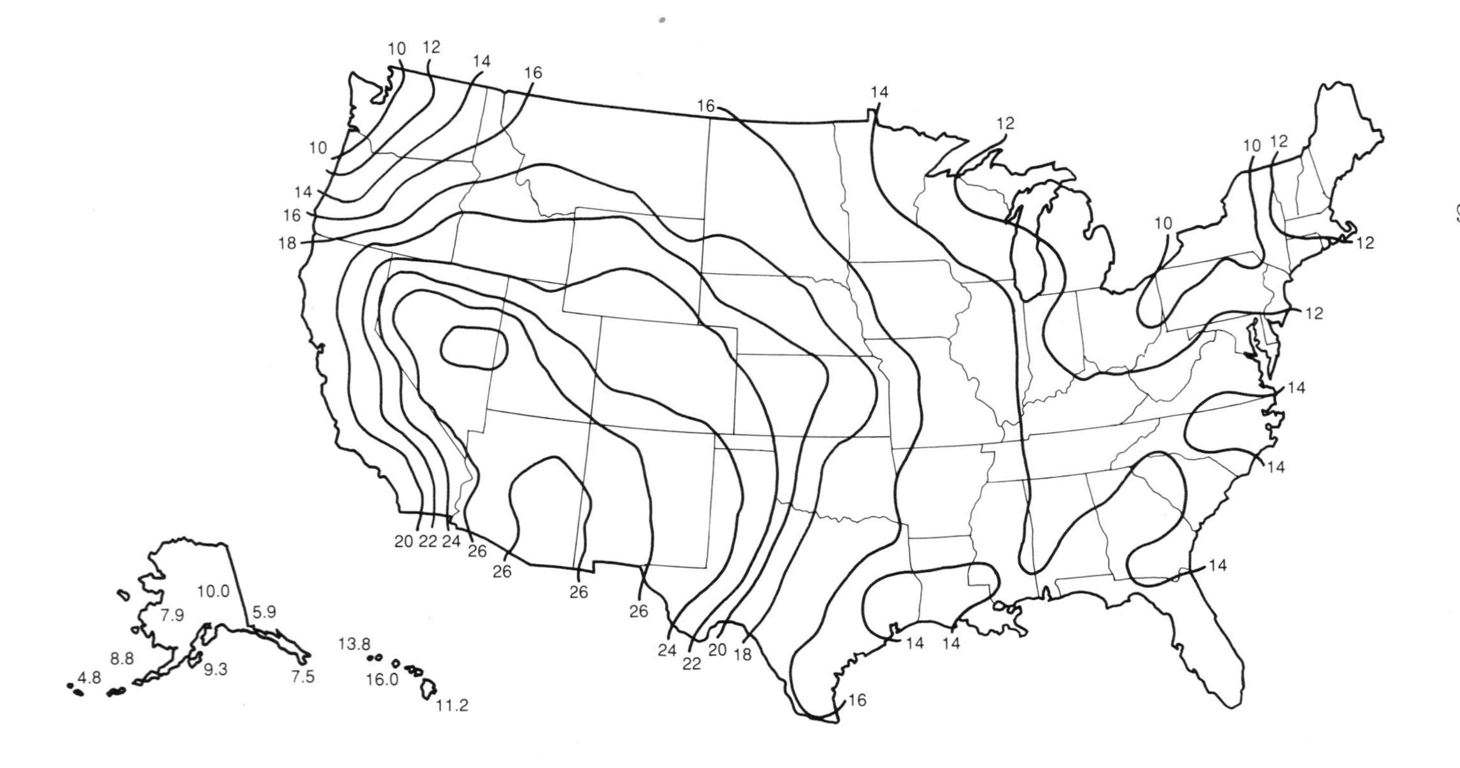
10
12
14
16
10
14
16
18
16
14
12
10
12
10
12
12
14
14
14
20
22
24
26
26
26
26
24
22
20
18
14
14
16
10.0
5.9
7.9
8.8
9.3
4.8
7.5
13.8
16.0
11.2

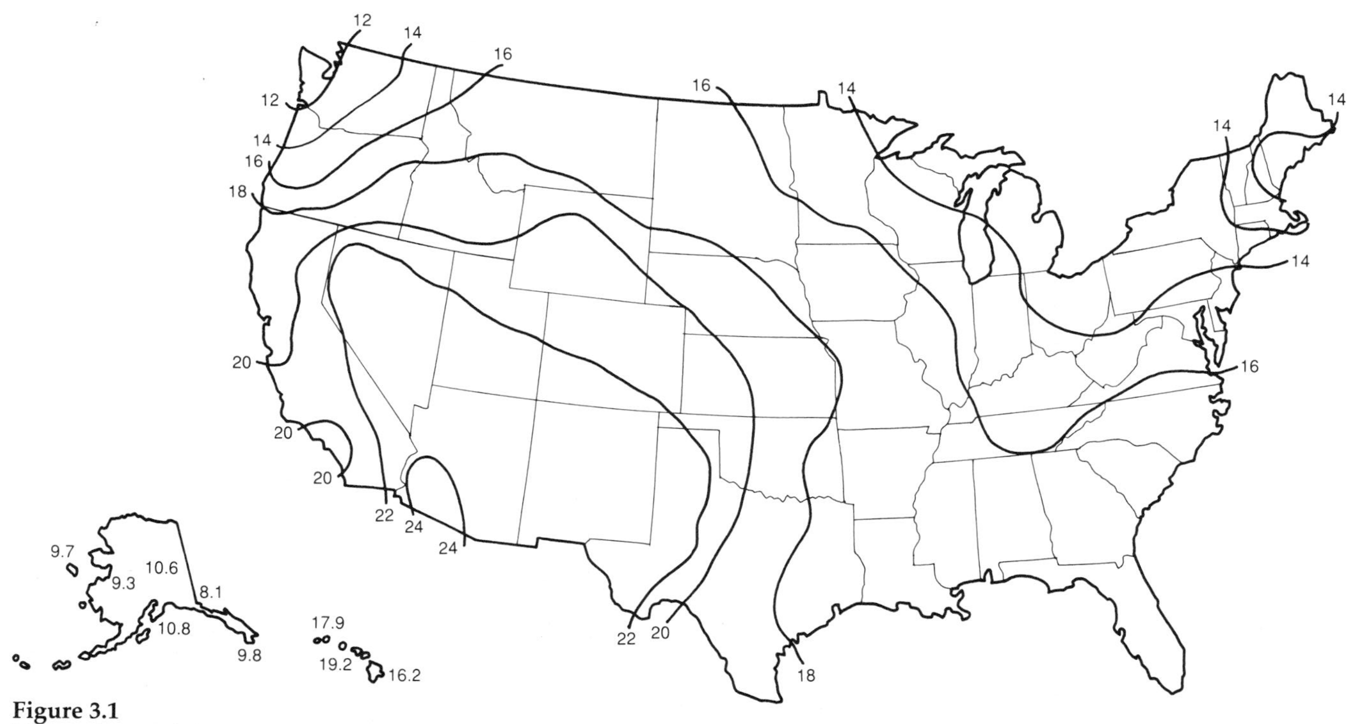

Figure 3.1
Maps of annual direct solar radiation on a surface normal to the sun's rays (top) and annual global (direct and indirect) solar radiation on a south-facing tilted surface (bottom), in megajoules per square meter per day ($MJ/m^2/day$). Note that the solar resource over one-third of the continental United States is as good as or better than that in southern California, where most solar development has taken place. Source: Hulstrom (1989).

one counts the energy used to generate electricity (EIA 1991b). Making extensive use of solar designs in new building construction — when coupled with efficiency measures such as insulation and low-emissivity (low-e) glass — could over the long term reduce this demand for fossil fuels by as much as 60 to 80 percent.

Passive Solar Design

Solar building systems are characterized as either passive or active. The passive approach uses a building's structure to capture sunlight and store and distribute heat, thus reducing the requirements for conventional heating and lighting. Large, south-facing windows that admit incoming sunlight but trap outgoing infrared radiation; tile floors, stone fireplaces, and brick interior walls that store heat from the sun and release it slowly at night; and simple architectural considerations such as orienting buildings to take maximum advantage of sunlight are all part of this strategy. Buildings can also be cooled by ventilating interior spaces, evaporating water, and using the earth to absorb heat, and windows, skylights, and interior spaces can be designed to optimize the use of natural light.

Figure 3.2
The Impact 2000 energy-efficient house in Brookline, Mass., collects solar heat through south-facing windows and produces hot water and electricity from roof-mounted collectors. Source: Solar Design Associates.

Besides saving energy, these design strategies can result in beautiful homes and buildings that are a pleasure to live and work in. Their cost depends on many factors, including the owner's taste, and varies with the degree of desired energy savings. Simple measures such as orienting a building toward the south to capture the winter sun, natural ventilation and sensible landscaping, increased south-facing window area for winter heating, and overhead shading for summer cooling can reduce a typical home's heating and cooling costs by 15-25 percent at little or no extra construction cost. Much of the reduction would come simply from moving windows from the north to the south side of the house. Adding thermal mass in the walls or floor while further expanding south-facing windows can result in overall energy savings of 30-70 percent. For custom-designed solar houses, the additional construction cost might be 10 percent, or perhaps $10,000 for a typical new home, but for tract builders working from standard designs, major solar savings could be achieved for much less (Saunders 1991, Aitken 1992).

These figures suggest that, on an annualized basis, the cost of energy saved could range anywhere from zero to $25 per gigajoule (GJ) depending on the particular design and builder, with a typical value (assuming standardized design and multiunit construction) of perhaps $5/GJ.[4] This compares very favorably with the current residential price of natural gas (also around $5/GJ).

Moreover, new technologies may further improve the cost effectiveness of passive solar design. Technologies under development include phase-change materials that store heat and moderate interior temperature more efficiently than water, rock, tile, and other ordinary materials; improved thermal glazings for windows and walls to reduce heat loss; light pipes and holographic film to bring sunlight deeper into interior spaces; and electrochromic window films that admit or block sunlight in response to a minute electrical current (DOE 1990b).

As might be expected in an energy system as complex as a building, passive solar design requires close attention to detail, and there are often complicated tradeoffs to consider. The Passive Solar Industries Council, among other groups, has developed helpful guidelines to aid architects and builders in resolving such issues. According to one rule of thumb, for example, the south-facing window area can equal up to 7 percent of a house's floor space without causing daytime overheating or wide temperature swings between day and night. If a larger window area — and hence larger solar contribution — is desired, then extra thermal mass, such as a tile floor or interior masonry wall, must be added according to simple formulas (PSIC 1990).

In general, commercial buildings present a more difficult challenge than residential buildings because the people, computers, lighting, and other equipment inside them act as internal heat sources. These internal loads can be so large that some buildings must be cooled year-round. In addition to more efficient lighting and equipment, shade trees, building orientation, and interior atriums can often do much to reduce cooling needs while making offices more attractive and pleasant places to work. Daylighting practices can be especially effective in cutting down on energy use for both lighting and cooling, provided that the building's lights are automatically controlled to dim in proportion to available daylight, so that interiors are not overlit. Direct solar heat gain is not useful in most commercial buildings, however, and indeed is often something to be avoided. Fortunately, it takes less than 1 percent of the sunlight incident on a building to meet its lighting needs, so that properly designed daylighting results in a smaller heat load than even the most efficient artificial lighting. This makes it possible to reduce the size of, or entirely eliminate, air conditioning systems.

Despite the general economic and aesthetic attractiveness of passive solar designs, they have not been adopted by the construction industry as widely as one might expect. About 250,000 passive solar homes have been built in the United States, most in the early 1980s, although the number of houses that incorporate at least some solar design features, such as skylights, is no doubt very much larger (DOE 1990b, English 1992). A lack of familiarity with solar design is one reason it has not widely caught on, but a more serious problem is that architects and construction firms have little direct incentive to incorporate solar features in the structures they build since they do not have to pay the costs of heating and cooling them later on. For this reason, appropriate building codes and incentive programs such as utility-sponsored rebates are essential for encouraging sensible energy design. Unfortunately, once a building is finished, the cost of retrofitting it with solar features often becomes prohibitive. With buildings, more than with almost any other type of energy investment, it is important to do things right at the outset.

Even so, there are signs of growing interest in passive solar design amongst buyers, architects, and builders. Some utilities, such as the Sacramento Municipal Utility District in California and Sierra Pacific

Power in Nevada, are developing programs to encourage passive solar design as an inexpensive means to reduce their customers' energy use (Osborne 1992, Bony 1992). Training workshops run by the Passive Solar Industries Council at various locations around the country have seen a surge in attendance, and federal agencies such as the National Park Service and General Services Administration are participating to find ways to incorporate passive solar design into their new buildings. A growing concern for the environment and a growing appreciation for the beauty of passive solar buildings are the most important factors driving this modest but encouraging trend (English 1992).

Solar Collectors

Solar collectors are discrete units that collect, store, and distribute solar energy for water heating, space heating, and space cooling. While most use pumps or fans, some manage all of this with no moving parts, relying instead on natural convective forces to circulate hot water or air through the collector, making them properly passive systems. Many types of solar collectors have been developed, the simplest and most popular of which is the flat-plate collector. In a typical solar water-heating design, a black metal plate absorbs sunlight and transfers the heat to pipes carrying water or, to prevent freezing in winter, a water-alcohol mixture. The absorber plate is topped by glass, and the rest of the system is surrounded by insulation. Over the years, this simple design has been improved so that, on average, about 50 percent of incident solar energy is transferred to the water; peak instantaneous efficiencies run as high as 80 percent (IEA 1987). Hot water systems have also become more reliable. In the late seventies, system owners reported an average of 2.5 significant problems per unit per year; by 1988 the rate was down to about 0.15 problems per unit per year (DOE 1990b).

In water- and space-heating applications, heat from the solar collectors is usually transferred to an insulated storage medium such as a water tank and distributed as needed. An innovative example is the Copper Cricket, manufactured by Sage Advance Corp. in Eugene, Oregon. This flat-plate collector relies on a "thermo-siphon effect" to pump an alcohol mixture from the rooftop collector to a tank as much as 36 feet below and back up again. Sunlight striking the collector boils the fluid, and the bubbles push the fluid above them to the top of the collector, where it drops down to a heat exchanger coupled to a storage tank. This flow, in turn, pushes the fluid that has already passed through the heat exchanger back up to the collector to repeat the cycle. According to the manufacturer, the three-square-meter, one-size-fits-all Copper Cricket can produce 90 percent of the year-round hot water needs for a typical family in Phoenix, and about 40 percent of the needs of a family in Boston.

Solar heat can also drive a cooling system, usually an absorption chiller (which operates on the same principle as a gas-powered refrigerator) or a desiccant-evaporator. Though some designs are effective, they are generally expensive and only practical on a large scale. Their performance could improve with the use of evacuated-tube collectors, in which the insulation of an ordinary flat-plate collector is replaced with a vacuum to minimize heat losses, or with concentrating collectors (such as parabolic troughs, discussed in the next section), both of which generate higher temperatures than flat-plate collectors; more effective desiccant materials are also under development (DOE 1990b). Overall, however, active solar cooling technologies do not appear very promising, especially when compared to much simpler and less expensive passive techniques.

The U.S. market for solar collectors has had a roller-coaster ride, growing rapidly in the early 1980s and then collapsing just as quickly with the decline in fossil fuel prices and the end of residential renewable energy tax credits in 1985. In 1984, some 250 manufacturers sold about 1.5 million square meters (16 million square feet) of collector (both flat-plate and concentrating). Within three years only about 60 manufacturers remained and sales had plunged to about 0.4 million square meters (4 million square feet) per year. Since then annual sales have leveled off, and in fact in 1989 (the last year for which data are available) there was a slight increase to about 0.6 million square meters (6 million square feet). Even at the market's peak, however, solar collectors never played a significant role in the nation's energy supply. Most of the more than one million systems that were sold were for heating hot tubs and swimming pools (EIA 1991b).

This experience has been echoed in Europe and other parts of the world, with some notable exceptions. In Cyprus, more than 90 percent of homes are now equipped with solar water heaters, as are at least 65 percent of homes in Israel. Israeli law now requires all new residential buildings smaller than 10 stories to use solar water heating. In parts of Australia, the market penetration is above 30 percent (Shea 1988, IEA 1987).

Two important factors limiting sales in the United States are the relatively high initial cost of active solar systems — ranging from $2,500 for small water-heating systems such as the Copper Cricket to $8,000 and up for residential space-heating and space-cooling systems — and the lack of long-term financing or leasing options for homeowners. Depending on the location, a solar water-heating system can be expected to save a homeowner about $250-$500 in energy costs per year if the fuel displaced is electricity, and about a quarter as much if the fuel displaced is natural gas.[5] Thus, if the owner invests his or her own money up front, it will not begin turning a profit for at least five years (assuming energy prices do not rise significantly in that time). But if long-term financing or leasing is available, then solar water heating can put money directly into the owner's pocket without any up-front investment. Suppose, for example, that the initial cost could be folded into a 30-year home mortgage.[6] Then the annual financing charge for a $2,500 system would be about $250, equal to or less than the energy savings if electricity is the competing fuel. Alternatively, electric and gas utilities could lease solar water heaters to their customers, just as they lease conventional water heaters.

Why are solar collectors so expensive? Not because they are complicated or difficult to manufacture; nothing could be simpler to make than a flat-plate collector, which consists of little more than sheet metal, copper tubing, glass, and insulation. The main reason is the lack of a secure market, which forces manufacturers to produce collectors at an inefficient rate and to charge high overhead costs. Sage Advance, maker of the Copper Cricket, manufactures on average about 100 units a month, hardly enough to justify mass production. At ten times this production rate, the company predicts the price could drop immediately by about one-third (Haines 1991).

Space heating accounts for only a small fraction — 10 percent or less — of all collectors sold. Here the problem is not just the initial cost of the system, but the seasonal mismatch between insolation (INcoming SOLar radiATION, measured as incident energy per unit area striking a horizontal surface) and demand. Whereas hot water is used all year, space heating is needed only during cold months, when sunlight is weak. The cost effectiveness of space heating could be greatly improved by storing summer heat for winter use (seasonal storage), a technique already being applied in Sweden. One such system, planned for the University of Massachusetts at Amherst, will supply space heating at a constant dollar levelized cost of about $11/GJ, approximately twice the current residential price of natural gas. (For a more complete discussion of seasonal heat storage, see chapter 8.)

Solar-Thermal Concentrating Systems

Buildings are heated to about 20°C (68°F) and use water at a temperature of no more than 80°C, but some types of solar collector can concentrate sunlight to the much higher temperatures needed for industrial applications and for generating electricity. The potential market for solar-thermal concentrating systems is consequently very large: Industrial use of fossil fuels amounts to about one-quarter of U.S. energy consumption, and electric utilities account for an additional one-third. Significant inroads have recently been made in the electricity market. From 1984 to 1990, over 350 MW of solar-thermal electric capacity were installed with private capital in California. Industrial process heat applications remain largely undeveloped, however.

There are three main types of solar-thermal concentrating collector: the parabolic trough, parabolic dish, and central receiver.[7] The parabolic trough is a curved reflector that focuses sunlight onto a line receiver, usually a vacuum-enclosed metal or glass pipe, in which a fluid is heated to temperatures as high as 400°C (750°F). These collectors are usually mounted on a north-south axis and track the sun. A parabolic dish consists of a bowl-shaped reflector that focuses sunlight onto a small receiver through which passes a heat-transfer fluid. Because of the higher concentration of sunlight, some of these achieve temperatures as high as 3,000°C (5,000°F), although generally they are designed to work at around 1,000°C (1,800°F). Dishes are usually mounted on a two-axis sun tracker. In a central receiver system, hundreds or thousands of sun-tracking mirrors, called heliostats, focus sunlight onto a large receiver mounted on a tower, where temperatures can reach well above 1,000°C.

Parabolic Troughs

Parabolic trough systems are well suited to producing heat for industrial processes, approximately half of which require temperatures less than 300°C. Several solar process-heat plants were built in the early 1980s for such diverse customers as a tractor company, a commercial laundry, a brewery, and a dyeing factory. More recently, one company, Industrial Solar Technology, has sold trough systems to state prisons in Tehachapi,

Figure 3.3
Solar collectors belonging to one of the nine LUZ solar-thermal power plants in southern California. The parabolic troughs focus sunlight onto a vacuum-enclosed pipe carrying a heat-transfer fluid, which is pumped to a central facility where it generates steam and electricity. Source: Union of Concerned Scientists.

California, and Denver, Colorado. These systems cost roughly $250 per square meter of collector, which in Denver implies an annual energy cost of around $6.50/GJ, depending on the cost of capital[8]— only slightly above the 1990 average price of natural gas for commercial users of about $5/GJ. While the current price of industrial gas is only about $3/GJ, gas price increases over the 20-year lifetime of the solar system should make this an attractive investment. Because of the need to look toward the long term, however, few private companies are inclined to invest in solar-thermal systems, and thus, barring new incentives, the main market for them is likely to be public institutions such as prisons, parks, and hospitals.

The cost of parabolic trough systems for industrial applications has declined sharply since the early 1980s, and their performance has improved steadily. Depending on the output temperature, the best technology today achieves up to twice the efficiency of systems developed a decade ago (Cleveland and Herendeen 1989, DOE 1990c). Nevertheless, there is still plenty of room for improvement. As is the case with flat-plate collectors, much of the price of today's concentrating systems goes to marketing and overhead. With a more secure market, Industrial Solar Technology estimates that it could cut the price of its systems to $150 per square meter, a 40 percent reduction. This would bring the levelized cost of energy in a climate like Denver's down to around $3.90/GJ.

By far the most successful use of parabolic trough technology to date has been for generating electricity. LUZ International, Ltd., built nine Solar Electric Generating Systems (SEGS) in California's Mojave Desert, the first in 1984 and the last in 1990. Collectively, these plants — now owned by private investors — generate about 95 percent of the world's solar-thermal electricity. The first SEGS was a 14 MW plant, the next six were 30 MW plants, and the last two were 80 MW plants, for a total of 354 MW. All are based on the same design: Sunlight strikes row upon row of parabolic troughs, heating a fluid inside the line receiver, a glass-enclosed pipe that runs through the center of each trough. This fluid is pumped to a central plant where steam and electricity are generated. The most recent plants cost about $3,000 per kilowatt, and according to LUZ, the levelized cost of electricity produced is around 10¢/kWh (Lotker 1991).[9]

Several factors were responsible for the company's remarkable success. First, California's aggressive implementation of the 1978 Public Utilities Regulatory Policy Act (PURPA),[10] combined with 25 percent federal and 25 percent state tax credits (yielding a net after-tax benefit of 38.5 percent), allowed the company to secure a market for its first plants even at their high initial cost. Second, LUZ succeeded in greatly reducing capital costs and improving performance for each successive plant it built. The installed cost fell from $5,979/kW for SEGS I to $3,011/kW for SEGS IX, a drop of 50 percent. In addition, the newer LUZ plants produce electricity 20 percent more efficiently than the older ones, thanks to higher temperatures achieved in the collector fluid and improved turbine efficiencies (Kearney et al. 1991).

Third, the LUZ plants have proved well suited to meeting peak summer daytime needs in the Los Angeles area, for which the local utility, Southern California Edison, is willing to pay a large premium. Even though only 40 percent of annual SEGS output is generated in peak and summer mid-peak periods, between 61 and 78 percent of annual revenues are earned then, depending on utility contracts (Kearney et al. 1991). Furthermore, since natural gas is used to supplement solar energy — up to the 25 percent allowed by PURPA for Qualifying Facilities — the nine plants are able to generate power reliably when its value is highest.

Although at least four other 80 MW LUZ plants were planned, and although the nine completed plants continue to operate, the company tragically went bankrupt in 1991, putting a halt to new construction, marketing, and research and development. LUZ's experience illustrates some of the great difficulties faced by renewable energy developers — even those with a proven track record. A major cause of the bankruptcy was uncertainty over available tax relief, a vital ingredient since under existing tax codes solar power plants must pay much higher taxes than conventional power plants of a similar size.[11] Because the federal energy tax credit has been renewed on an annual basis since 1985, LUZ faced an artificially imposed, year-end deadline for completing its tenth plant. Before starting construction, however, the company needed assurance that the new plant would be exempt from state property taxes on the land occupied by the collector field. Unfortunately, legislation to provide this exemption, which had been approved in previous years, was held up several months by California's new governor, leaving LUZ with barely 7 months to build the plant. Faced with higher construction costs and risks, LUZ was unable to secure funding for the plant and was consequently forced into bankruptcy (Lotker 1991).

Size limitations imposed by PURPA also played an important part. Until 1987, solar plants were restricted by PURPA to no more than 30 MW capacity. This restriction significantly raised the cost of the first seven SEGS plants, as demonstrated by the fact that when the size limit was raised to 80 MW in 1987, the cost of energy from the SEGS VIII and IX plants dropped by about one-third. The size limit was finally eliminated in 1990, but not in time for LUZ to begin construction of larger plants that would have achieved even lower costs (Lotker 1991).

Despite LUZ's surprising successes, it is a matter of some debate whether the parabolic trough technology is truly a viable option for electricity generation in the long term. The Department of Energy, for one, has in recent years favored parabolic dish and central receiver systems because of their higher inherent thermal-electric efficiency (a result of the higher temperatures they generate), although the very fact that LUZ got as far as it did with parabolic troughs suggests that they have some important advantages (namely, they are well suited to low-tech assembly-line construction of very large units). Unfortunately, the question may never be settled. Not only did LUZ go bankrupt before it could build any plants larger than 80 MW, it was also prevented from developing the next generation technology that might have allowed it to survive entirely without tax credits or exemptions. Plans called for generating steam directly in the receivers at potentially higher efficiency and lower cost than in the current scheme.

Parabolic Dishes

Parabolic dish systems produce higher temperatures than troughs, making them suitable for such industrial applications as the production of metals, glass, cement, paper, and brick, as well as for generating electricity at higher efficiencies than are possible with trough systems. Most multidish system designs are conceptually similar to the LUZ SEGS design: A heat-transfer fluid is passed through receivers positioned at the focal point of the dishes, then piped to a central location for use.

In contrast to parabolic trough systems, dish systems have not yet found a significant market. In part this may be because the high-temperature industrial heat market is more specialized and demanding than the medium-temperature market for which trough systems are designed. It may also be that the technology is somewhat more complex than the trough technology (requiring a two-axis sun tracker) and less amenable to economies of scale. Whatever the reason, most of the dish systems that have been built were sold to governments and utilities for individual testing, and only a few have been adapted for commercial energy production. The two largest systems, a 5 MW electric plant built in Warner Springs, California, by LaJet Incorporated, and a 3 MW (thermal) facility operated by Georgia Power Company in Shenandoah, Georgia, were shut down in the late 1980s.

In recent years, as federal R&D funding has declined, most research in dish technology has concentrated on developing advanced components, such as light-weight stretched-membrane reflectors and high-efficiency receivers, rather than complete systems. The exception is the dish-Stirling concept, in which each parabolic dish is equipped with its own electricity generator, a Stirling-cycle engine. This concept may be well suited for small-scale, remote applications, particularly in developing countries, as well as for utility generation. One model under development by the Department of Energy and Cummins Power Generation company is designed to produce 5 kilowatts, a good size for a small rural village. In this model, a stretched-membrane parabolic dish concentrates sunlight onto a receiver filled with sodium. The heat vaporizes the metal, and the vapor then condenses on the heater head of the attached Stirling-cycle engine, causing the helium gas inside the engine to expand. This expansion (and subsequent contraction) of the helium drives a piston and an alternator to generate electricity. An earlier dish-Stirling prototype achieved a net solar-electric conversion efficiency of 29 percent, the highest so far achieved by any solar-thermal technology (IEA 1987, SNL 1991a).

Current Department of Energy plans call for these systems to be produced commercially by 1995 or 1996 at a cost of about $3,000/kW. At this (probably optimistic) price, they could be quite competitive with diesel-generated electricity used in many remote applications. Larger dish-Stirling modules in the 25-30 kW range appropriate for utility use could ultimately cost perhaps $1,000-$1,500/kW. A key requirement for successful marketing of this technology will be to demonstrate long-term reliability and low maintenance costs. Early dish-Stirling models were based on automotive engines, which are designed to run only a few thousand hours between overhauls. New engines must be built to last at least ten times as long.

Central Receivers

Central receiver systems were the initial focus of solar-thermal research in the United States, and they continue to dominate the Department of Energy solar-thermal program. Several experimental plants were built in the early 1980s in France, Italy, Spain, Israel, Japan, the Soviet Union, and the United States. The largest of these, the 10 MW Solar One plant, is located near Barstow, California, in the Mojave Desert, a short distance from the LUZ installations. As was the case with the dish technology, the drop in federal R&D funding in the eighties led researchers to concentrate on improving basic system components, such as heliostats (mirrors), receivers, and thermal transport and storage systems, rather than building new facilities. Solar One was closed in 1988 after operating successfully for six years.

The recent increase in funding for solar energy has led to renewed work on central receiver systems, however. A utility consortium, which is expected to be supported partly by federal funds (not yet approved), plans to refurbish Solar One and restore it to operation by 1994 (Public Power Weekly 1991). "Solar Two" represents the second generation of central receiver technology. The first generation used heat to turn water directly into steam and stored any excess energy in a medium-temperature oil and rock bed. Among other problems, the steam pressure dropped quickly when clouds blocked the sun, forcing the power plant to shut down, and it took time to restart once sunlight returned. Also, there were substantial efficiency losses in transferring heat from the steam to oil and back to steam. Solar Two will use a mixture of molten sodium-nitrate and potassium-nitrate salts heated to nearly 1,000°C for both heat transfer and thermal storage, an approach pioneered in France. Some of the molten mixture will be used immediately to generate steam for electricity production; the rest will be stored in tanks for use after sunset or during cloudy intervals. In theory, such a system could provide continuous power on a 24-hour basis, but Solar Two will be designed to shift daytime output only a few hours into the evening so as to match Southern California Edison's peak loads as closely as possible (SNL 1991b). The refurbished facility will also test stretched-membrane heliostats in a portion of the reflecting field.

Using Solar Two's advanced technology in much larger plants could conceivably make central receivers competitive with conventional alternatives for generating intermediate and peak power. A cooperative study by two utilities (Pacific Gas and Electric and Arizona Public Service) and two engineering consulting firms (Bechtel and Black and Veatch) concluded that advanced central receivers of 200 MW size could produce electricity at a levelized cost (under conventional utility financial assumptions) of 8.4¢/kWh (Hillesland 1989, converted to 1990 dollars).[12] Possibly even lower costs could be achieved with more advanced technology such as direct-absorption receivers. Utilities will have to make a large initial investment to reach this level of performance, however, making the commercial prospects of this technology highly uncertain. Merely refurbishing and modernizing the Solar One facility will cost an estimated $39 million, or $3,900 per kilowatt. Assuming $3,000 per kilowatt for the first commercial 200 MW facility implies an up-front investment cost of $600 million — a large pill for any utility to swallow.

The high financial hurdle that must be crossed to make central receivers economic illustrates a basic problem shared to varying degrees by all solar-thermal electric technologies (with the possible exception of the dish-Stirling concept). At the modest scales of, say, the 10 MW Solar One or 30 MW LUZ SEGS, the plants are simply too expensive to compete in today's electricity markets without special tax incentives or other supports.[13] Yet without the experience of building and operating these smaller plants, few utilities or private companies will be willing to risk investing in a much larger plant. LUZ came closest to overcoming this barrier by taking advantage of the favorable utility contracts available in California and piecing together financing for successively larger plants, but with the failure of this enterprise it is difficult to see where the next

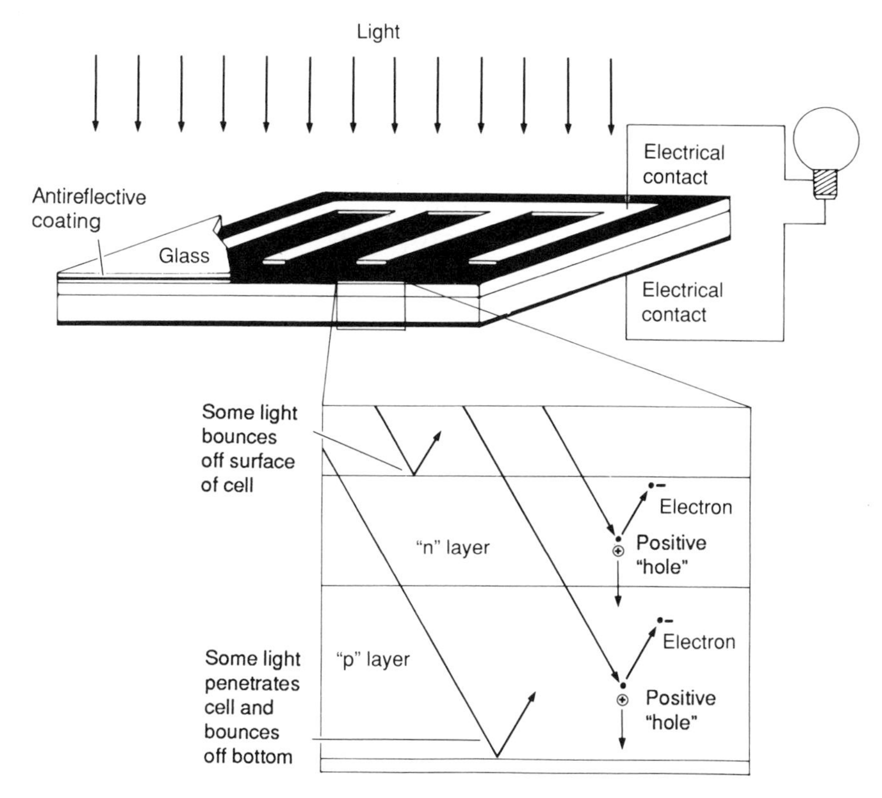

Figure 3.4
Schematic of a typical photovoltaic cell. An electric field is created at the junction of two dissimilar semiconductor materials. When electrons are knocked loose from atoms by incident light, the negative electrons and positive "holes" are swept onto opposing electrical contacts. When the circuit is completed, a current is established.

opportunities for commercial solar-thermal electric development will emerge. Fortunately, this is not a difficulty shared by photovoltaic systems, which, as we will see, can be scaled to suit almost any application with little impact on cost.

Photovoltaic Cells

Photovoltaic cells are the most elegantly simple of all solar-electric technologies, and perhaps for this reason have received the most public attention. Initially developed to power satellites in space, these devices have proven extraordinarily useful in a wide range of applications, from tiny solar cells for wristwatches and calculators to kilowatt-scale systems for remote and specialized power applications. The photovoltaic industry, while still small at 50 MW of annual sales worldwide, is growing at a rate of 20 percent per year. Substantial cost reductions and performance improvements will be needed before photovoltaic electricity begins to penetrate the bulk electricity market, but there is every reason to think this goal can be met, if sufficient investments are made in research and development.

Photovoltaic cells generate electricity when photons of light knock electrons loose from atoms. The negatively charged electrons and positively charged "holes" are swept onto opposing metal contacts by a voltage created between two semiconductor materials. Closing the circuit establishes an electric current. Individual cells generate 0.6 to 1.2 volts each and are wired in parallel or in series to provide greater current or voltage, respectively. Since 1954, when the first modern cell was demonstrated, a vast array of materials and designs has emerged from government and private research, and new developments are occurring all the time.

Silicon (Si) has long been the dominant material used to make photovoltaic cells because of its familiar properties, availability, and wide use as a semiconductor in computer chips. Single-crystal silicon cells were the first to be developed and still hold just under half of the market, though other silicon technologies are also being used. Among silicon cells, those made from a single crystal offer the highest conversion efficiency from sunlight to electricity. This is a key factor in cost, since the more efficient a cell, the more electricity it can produce from a smaller collector area. In 1988, researchers set a new record when they measured an efficiency of 22.8 percent in a small single-crystal cell under normal, unconcentrated light. Commercially available flat-plate modules (integral units with many individual cells) currently demonstrate an efficiency of about 15 percent (DOE 1991b).

The main disadvantage of single-crystal silicon cells is their cost. The cells must be fairly thick (100-300 micrometers, or millionths of a meter) to absorb sunlight of the proper wavelengths. Not only do they require a lot of material, but growing single crystals and then cutting them into wafers is a slow, wasteful process. Some innovative ideas have been proposed that may greatly reduce the material requirements, however.

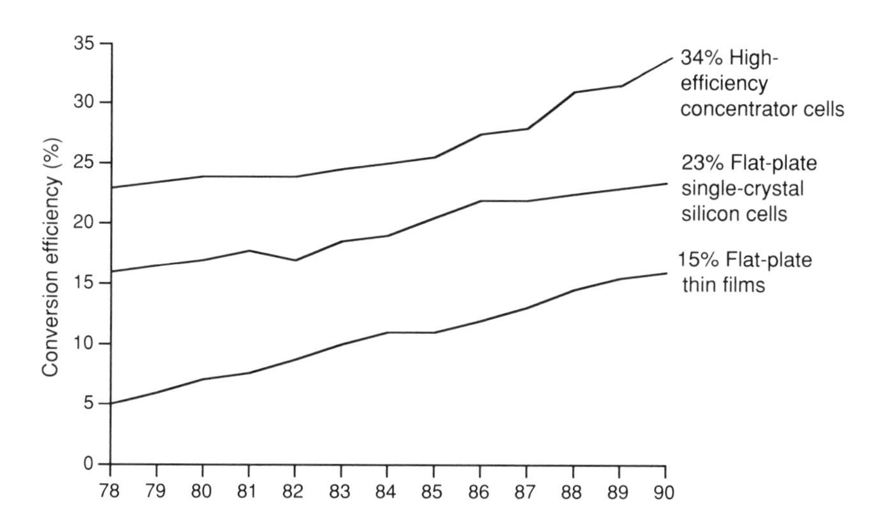

Figure 3.5
The efficiencies of photovoltaic cells developed and tested in the laboratory have risen steadily since the 1970s, though commercial modules still lag behind. Source: DOE (1991b).

In a process still in early development, for example, Texas Instruments is growing tiny spheres of single-crystal silicon, which are then shaved flat on opposing surfaces and mounted in dense numbers on conducting films (Levine, Hotchkiss, and Hammerbacher 1991). The films provide structural strength and flexibility and serve to conduct the electricity generated. A partnership of Texas Instruments and Southern California Edison has announced plans to market these so-called "spheral" cells by the mid-nineties.

Another approach to making single-crystal silicon cells less costly involves adding lenses or reflectors to concentrate sunlight onto a smaller cell area. The higher cost of the concentrating collector and sun-tracking system it requires is offset by the lower cost of the cell itself. What is more, the efficiency of cells under concentrated light is higher than under normal light. The champion silicon concentrating cell achieved an efficiency of 28.2 percent under sunlight concentrated 100 times.

An alternative to single-crystal silicon is polycrystalline silicon, which has somewhat lower efficiency but costs less to manufacture. In one production method, molten silicon is drawn into sheets (called ribbons), while in another, it is cast as ingots and sliced into wafers. The highest efficiencies measured for individual cells are in the range of 15-17 percent under normal light, which is considerably lower than the efficiencies of single-crystal cells. The decrease in efficiency is partially offset by better use of space and less material waste, however, since polycrystalline cells are square and can be more tightly packed onto a flat-plate module than circular single-crystal wafers.

A more radical departure from high-efficiency, high-cost single-crystal cells are thin films made from amorphous, or noncrystalline, silicon (a-Si). Since amorphous silicon absorbs light easily, the films can be very thin, just one or two micrometers thick. Moreover, these thin films can be deposited on various inexpensive substrates, such as glass or metal, by techniques easily adapted for mass production. One drawback is their low efficiency, about 5-7 percent for today's modules, although the best small-area cells today have reached efficiencies of more than 13 percent (Luft, Stafford, and von Roedern 1991). In addition, when amorphous silicon films are first exposed to sunlight, their efficiency drops about 15 to 30 percent before stabilizing. Research sponsored by the Department of Energy is aimed at producing amorphous silicon modules with stabilized efficiencies of 10 to 12 percent. Companies have encountered unexpected problems in manufacturing reliable a-Si modules in large quantities, however, so the future of this technology for power applications is somewhat in doubt. Currently their main application is consumer products.

The various problems with silicon have heightened interest in making thin-film photovoltaic cells from nonsilicon materials. Gallium-arsenide (GaAs) cells are the most efficient, achieving 24.2 percent efficiency under normal light and 29.2 percent efficiency under concentrated light in the laboratory. They are expensive and difficult to manufacture in large quantities, however. A more important advance has been the development of copper-indium-diselenide ($CuInSe_2$) and cadmium- telluride (CdTe) thin films with potential practical efficiencies of 16 and 18 percent, respectively, using existing designs (Ullal et al. 1991). Other experimental cells, such as one that uses inexpensive semiconductor materials sensitized with a light-harvesting dye, are also under investigation (O'Regan and Gratzel 1991).

The cell designs mentioned so far all consist of two semiconductor layers that meet at a single junction. Adding more layers of different material can raise the efficiency by tapping a broader spectrum of sunlight; these are known as multijunction cells. One company, Boeing, has achieved a remarkable efficiency of 34.2 percent using a multijunction gallium-arsenide alloy cell under concentrated light. Multijunction amor-

Photovoltaic Systems for Home Use

For consumers interested in providing their own off-grid power, the *Alternative Energy Sourcebook* (Real Goods 1991) lists everything from small wind turbines to large photovoltaic systems and electric automobiles. A $2,795 remote home PV kit capable of providing 720 Wh per day, for example, will operate lighting, 110 volt television and video-cassette recorder, satellite dish receiver, computer, limited water pumping, sewing machine, drill, small vacuum, small microwave, juicer, blender, and other appliances for up to 30 minutes. It comes complete with:

- four 48-watt PV modules
- one latitude adjustable four-panel mount
- one Trace C-30A charge controller
- one 30A two-pole fused disconnect with fuses
- one UL-listed DC load center with two 15A breakers
- two 15A circuit breakers
- one Two Channel Digital Volt-Amp Monitor
- four 6V, 220 amphour deep cycle batteries
- one Trace 612 12V to 110V power inverter
- one Remote Home Kit Owner's Manual

Photovoltaic power system kits for mobile homes and boats, as well as free-standing photovoltaic modules and trackers, are also listed and described.

phous-silicon cells may be a solution to that material's relatively low efficiency and unstable performance. While the maximum theoretical efficiency of photovoltaic cells is limited by physics and differs from material to material, there is still considerable room for further improvement in laboratory efficiencies, and commercial modules lag 5-10 years behind the state of the art.

Along with improvements in the performance of photovoltaic cells has come decreasing cost. In 1976, the price of photovoltaic modules averaged $44,000 per peak kilowatt of capacity (1986 dollars), but by 1986

Figure 3.6
This 2 MW photovoltaic facility is located next to the Rancho Seco nuclear power plant, which was shut down in 1990 by public referendum. Source: Department of Energy.

the price had dropped to just over $5,000 per peak kilowatt, and the current price is estimated to be $4,000-$5,000 per peak kilowatt (Shea 1988, Taschini and Iannucci 1991).[14] Other "balance of system" components, such as power conditioners, direct-to-alternating current converters, and controls, may add $1,000-$2,000 per peak kilowatt to the cost of large (multikilowatt) systems, and more if the systems are smaller (Jones 1992).[15]

Although the progress has been impressive, the price is not as low as once predicted. In 1979, the Department of Energy set cost goals for photovoltaic cells of $4,400 per kilowatt by 1982, $1,200 per kilowatt by 1986, and $250-$700 per kilowatt by 1990 (Kendall and Nadis 1980, converted to 1990 dollars). A decrease of more than 80 percent in government funding for photovoltaic research since the early eighties is one factor behind the failure to meet these early goals, though unexpected technical difficulties, such as light degradation in amorphous silicon cells, have also played a part. In addition, strong market demand has encouraged manufacturers to keep prices high in recent years, despite decreasing production costs, and this trend may continue for a few years before competition forces prices down again (Boes 1992).

With the drop in cost in the eighties came new markets for photovoltaic cells. From 1976 to 1983, annual world shipments of photovoltaic modules grew from 0.5 MW to 20 MW. After 1983, photovoltaic sales held steady for a few years, but since 1988 they have grown rapidly, reaching 55 MW in 1991 (EPRI 1991, DOE 1991b, Maycock 1992a). This growth has been accompanied by significant changes in the nature and makeup of the photovoltaic industry. Single-crystal silicon cells initially dominated sales, but their share has gradually declined as polycrystalline and amorphous silicon entered the market. At the same time, the U.S. industry's once-overwhelming share of the world market has fallen to roughly one-third, in large part because Japanese companies moved aggressively to capture the market for amorphous silicon cells in consumer products. The U.S. industry has traditionally concentrated on crystalline silicon for power applications, although some manufacturers, such as Advanced Photovoltaic Systems (formerly Chronar) and Solarex, are producing amorphous silicon thin films.

The largest market for photovoltaic cells is currently in providing remote power for equipment and communities far from utility grids. Thousands of remote communications systems, water pumps for livestock, navigation aids, and warning beacons now rely on photovoltaic systems to provide power, thereby eliminating the need to recharge batteries or maintain and refuel diesel-electric generators. In the United States and Europe, photovoltaic systems are frequently installed in vacation homes with typical units of a few hundred watts capacity costing $2,000-$5,000. The total potential market for such remote systems, according to one estimate, is a few tens of gigawatts, representing several tens of billions of dollars at current prices (IEA 1987).

Providing power for villages in less-developed countries is a fast-growing market with very large potential. The United Nations estimates that over two million villages worldwide are without electric power for water supply, refrigeration, lighting, and other basic needs. The cost of extending utility grids to such areas is often prohibitive, and remote photovoltaic systems could meet much of that need at lower cost. Once photovoltaic module prices reach $2,000-2,500 per peak kilowatt, they could virtually replace diesel-electric generators in this market (Taschini and Iannucci 1991).

In the United States and other industrialized countries, the most important new markets for photovoltaic systems in the near term will probably be medium-sized (100-500 kW) power stations designed to reduce peak demand for conventional power and smaller systems for grid-connected homes and businesses. At the moment, photovoltaic systems are still too expensive for most such applications. The current cost of photovoltaic electricity (using conventional utility financial assumptions) is in the range of 25-50¢/kWh, depending on the size of the system, or about five to ten times the cost of electricity generated from conventional baseload power plants and two to five times the cost from conventional peaking power plants.

Nevertheless, even at these costs, "high value" utility applications are beginning to emerge. For example, one utility, Pacific Gas and Electric in northern California, has found that photovoltaic systems are cost effective today in areas where they serve to reduce peak loads on transmission and distribution (T&D) networks that must otherwise be upgraded at considerable cost. In several parts of PG&E's service territory, for example, the marginal cost of upgrading the T&D network exceeds $6,000/kW, making photovoltaic systems a potentially attractive investment. A 500 kW photovoltaic facility is being installed near Fresno, California, to test this concept, and other utilities are moving to follow PG&E's example (Shugar 1991, Maycock 1992b).

If the Department of Energy's goals prove to be realistic, then photovoltaic system costs may drop to $3,000/kW by the mid-1990s, becoming competitive for meeting utilities' peak summer power needs, and possibly to $1,000/kW by the turn of the century, when these systems could begin to compete in the bulk electricity market. The three key requirements for achieving lower costs will be producing reliable commercial modules with efficiencies approaching those attained in the laboratory, manufacturing photovoltaic systems in much larger quantities and using less expensive production methods, and reducing the cost of support structures, wiring, power converters, and other balance-of-system components.

Of these three requirements, increasing the scale of production may be the most easily achieved and most effective in lowering overall system costs. A review of manufacturers' estimates and engineering studies suggests that the cost of amorphous-silicon photovoltaic modules could drop to $1,000/kW simply with the expansion of individual factory production capacities from the present 1 MW per year to 10 MW per year (Carlson 1989). The Department of Energy estimates that prices will fall 70 percent for every tenfold increase in production experience (cumulative megawatts produced over the years) and 40 percent for every tenfold increase in factory production capacity (DOE 1991b). The cost of power processing components, now specialty items because of the relatively low volume of PV sales, could drop from the present $300 to $500 per peak kilowatt to just $100 per peak kilowatt if utility-scale units could be manufactured in quantities of 100 to 1,000 per year (Jones 1992).

Greater experience and familiarity with photovoltaic systems will be essential if electric utilities are to have confidence in this technology. Over a dozen modest demonstration facilities (greater than 100 kilowatts peak capacity) and numerous small ones have been built worldwide since 1981. The largest was a 6.4 MW demonstration plant using single-crystal silicon cells built in 1983 by ARCO Solar for PG&E in Carissa Plains, California, which has been dismantled and sold off. A 2 MW photovoltaic system that uses single-crystal flat-plate collectors has been providing the city of Sacramento, California, with power since 1986. Experience with such plants has shown that, as a rule, photovoltaic systems are well suited to utility applications: They produce consistent power that is well matched to daytime load profiles, their performance is well understood and predictable, failure rates of modules are low (0.5-1 percent or less per year), and operations and maintenance costs are minimal (about 0.5¢/kWh or less) (SNL 1987a, 1987b).

Although the number of demonstration projects declined during the 1980s, several are still in the works. The most important of these is Photovoltaics for Utility-Scale Applications (PVUSA), a program jointly funded by the federal government and electric utilities, led by PG&E. This ambitious project will compare and evaluate several photovoltaic pilot systems of about 20 kW size and also test utility-scale systems of 250-500 kW size under realistic operating conditions. Preliminary results show wide differences in the performance of systems now on the market, with some falling well short of their manufacturers' specifications. This type of effort gives both manufacturers and utilities valuable information about how well commercial products perform in the field and should motivate rapid improvements in the technology (Weinberg, Hester, and Townsend 1991).

Over the years, photovoltaic systems have also been installed in a number of homes, communities, and businesses connected to electric-utility grids. Several experimental community projects, such as one in Gardner, Massachusetts, that serves 35 homes, businesses, and town office buildings, have been established around the country. Ultimately such applications may prove one of the most important markets for photovoltaic technology. A key advantage of installing photovoltaic systems at the customer's end is that electricity prices are higher there than at the generating plant. Although photovoltaic systems of only a few kilowatts in size tend to be more expensive per kilowatt than larger systems, with standardization of design and installation, economies of scale are still possible. At the moment, however, electric utilities rarely charge residential consumers time-of-day rates that reflect the true cost of electricity generated during peak periods, and it is on these higher rates that residential photovoltaic systems will depend for their initial viability. Nevertheless, the long-term prospects for such applications appear bright, since photovoltaic systems are simple, require little or no maintenance, and pose no unmanageable safety hazards.

Environmental, Health, and Safety Issues

All energy technologies have some impact on the environment, and solar technologies are no exception. Compared to those of fossil-fuel technologies, however, their impacts are very few, and most or all can probably be controlled through proper regulation.

The large amount of land required for certain solar applications, such as central station power plants, may pose a problem in some areas, especially where wildlife protection is a concern. The LUZ solar-thermal power plants, for example, were built on desert land, but nevertheless it was necessary to take measures to preserve wildlife species. This problem is not unique to solar power plants: Generating electricity from coal actually requires as much or more land, per kilowatt-hour produced, than generating it from solar energy, if the land used in strip mining is taken into account (Meridian 1989). Solar-thermal power plants (like most conventional power plants) also require cooling water, which may be costly or scarce in desert areas, although this has not been a problem for LUZ.

Since solar systems generate no air pollution during operation (unless natural gas or another fuel is used as a backup), the primary environmental, health, and safety issues involve how they are manufactured, installed, and ultimately disposed of. Energy is required in the manufacture and installation of solar components (even the windows and thermal floors in a solar building), and any fossil fuels used for this purpose will generate pollutant emissions. Thus, an important question is how much fossil energy input (and what type) is required for solar systems compared to the fossil energy consumed by comparable conventional energy systems. This question is difficult to answer in general, as only a few studies have been done, and the results vary widely depending on assumptions such as climate, the state of technology, and other factors. Overall, however, it appears that the energy balance is favorable to solar systems in applications where they are cost effective, and it is improving with each successive generation of technology. According to some studies, for example, solar water heaters increase the amount of hot water generated per unit of fossil energy invested by at least a factor of two compared to natural gas water heating alone and by at least a factor of eight compared to electric water heating alone (Hall, Cleveland, and Kaufmann 1986). Solar-thermal concentrating systems, on the other hand, increase the heat return by a factor ranging from 1.8 to 10, depending on the sunniness of the location and the output temperature (Cleveland and Herendeen 1989).

Materials used in solar systems can create health and safety hazards for workers and anyone else coming into contact with them. In particular, the manufacturing of photovoltaic cells often requires hazardous materials such as arsenic and cadmium. Even relatively inert silicon can be hazardous to workers if it is breathed in as dust. Workers involved in manufacturing photovoltaic modules and components must consequently be protected from exposure to these materials. There is an additional, probably very small, danger that hazardous fumes released from photovoltaic modules attached to burning homes or buildings could injure firefighters (Holdren et al. 1983).

None of these potential hazards is much different in quality or magnitude from the innumerable hazards people face routinely in an industrial society like ours. Through effective regulation, the dangers can very likely be kept at a very low level. Perhaps the most serious concern for the future is that photovoltaic manufacturing will be exported to developing countries, where environmental, health, and safety regulation and enforcement may not be sufficient to protect the public.

Notes

1. Only geothermal and tidal energy are not derived in one way or another from sunlight.

2. Solar radiation data are from *CRC Handbook of Chemistry and Physics*, 67th Edition (Boca Raton, Fla.: CRC Press, 1986). Fossil-fuel reserves are assumed to be double proven oil and gas reserves and double estimated recoverable coal reserves; data are from EIA (1991b).

3. It has been proposed, for instance, that vast amounts of solar electricity could be generated in southwestern deserts and converted to hydrogen, which could then be piped to other parts of the country. Producing enough hydrogen to replace 1990 U.S. oil use — 34 EJ per year — would require some 64,000 square kilometers (24,000 square miles), about 0.5 percent of all U.S. land area and 7 percent of U.S. desert area (Ogden and Williams 1989). By comparison, as of 1982, about 30 percent of nonfederal land was classified as cropland and about 3 percent as urban and built-up land (Bureau of the Census 1990).

4. This calculation assumes a base construction cost of $6 per square meter ($60 per square foot) of floor space, base annual energy consumption for space heating of 300 MJ per square meter (30,000 Btu per square foot), and an annual inflation-adjusted cost of capital of 5 percent (roughly corresponding to current mortgage interest rates).

5. This range assumes a 4 square meter (40 square foot) collector operating at 50 percent average efficiency, an electricity price of 8¢/kWh ($23/GJ), and a natural gas price of $5/GJ.

6. Solar water heaters could be accepted as an energy-saving measure in home mortgages approved by the Federal Housing Administration if they were certified by a suitable agency such as the industry-operated Solar Ratings and Certification Corporation.

7. Two other solar-thermal approaches not addressed here are lens-concentrating systems and solar ponds. Neither appears likely at the moment to find widespread application. Solar ponds are discussed briefly in chapter 8.

8. This calculation assumes a heat output of 1.9 $GJ/m^2/yr$ (170,000 $Btu/ft^2/yr$) at a temperature of 150°C (300°F) and an inflation-adjusted annual cost of capital of 5 percent. At lower temperatures the system efficiency will be higher and the cost lower, whereas at higher temperatures the reverse will be true. Performance data are taken from Cleveland and Herendeen (1989) and assume Industrial Solar collectors with improved optical coatings.

9. This price reflects both the relatively low cost of capital available to LUZ because of its relatively high debt-to-equity ratio and federal and state tax credits. If the SEGS were utility-owned and financed, the cost of electricity would be around 12¢/kWh.

10. This federal law requires utilities to buy power from independent producers at a price determined by the "avoided cost" of generation, or the cost the utility would have to pay to generate the same electricity from other sources. California was one of the most aggressive states in interpreting and implementing this law. Among other steps, the state established 10-year fixed-price contracts (Interim Standard Offer 4, now unavailable) for PURPA Qualifying Facilities.

11. Property taxes, sales taxes, and payroll taxes all affect solar power plants to a disproportionate degree because of the large land area they occupy, the high cost of purchased equipment, and the need to employ people to clean and service the collector field. An 80 MW SEGS plant, for instance, would pay about

$26 million in sales and property taxes in California over its 30-year lifetime, whereas an 80 MW gas-fired power plant would pay just $7 million (Lotker 1991).

12. This is the cost of the fifth to the tenth commercial plants, including expected cost reductions from accumulated engineering experience. The cost of electricity for the first 100 MW commercial plant would be 11.4¢/kWh (Hillesland 1989).

13. This is not primarily the fault of the solar collectors but rather the fact that the plants must rely on steam turbines to generate electricity. The per-kilowatt cost of steam turbines strongly depends on size, and below about 100 MW they are usually not economic.

14. A peak kilowatt refers to the direct-current power generated at a nominal solar incidence of 1 kilowatt per square meter, roughly equivalent to noontime sun on a clear day. Thus, a one-square-meter, 10-percent efficient module would have a peak capacity of about 100 watts dc and today would cost $400-500 (not including balance-of-system costs).

15. The term balance of system is often used, incorrectly, to refer to all nonmodule costs, including both hardware and taxes, contingency fees, design costs, and other factors. These latter items can add 20 to 30 percent or more to the final cost of a utility-scale PV system but should become less important as utilities gain experience with this technology.

4 Wind Energy

Wind, one of the oldest energy sources known to humanity, has been used for millennia to pump water, thresh grain, and propel ships. In the early part of this century, windmills for pumping water were a common sight in rural areas of the United States, until rural electrification programs in the thirties and forties caused many of these reliable and durable machines to be abandoned (Metz and Hammond 1978).

Like the solar industry, the wind industry experienced a revival in the seventies and eighties. From 1981 to 1986, more than 15,000 utility-scale wind turbines with a total peak capacity of 1,300 MW were installed in California, where the vast majority of U.S. and world wind power development has taken place (Gipe 1991). Over the same period, about 5,500 small wind turbines were installed throughout the United States, mainly at rural homes (Bergey 1990).

This rapid development was far from smooth, however. The rush to build wind turbines brought many poorly designed machines to market that were either inefficient or unable to withstand the mechanical stresses caused by variable and high-speed winds; many were placed in less-than-optimal locations. Coupled with these technical problems, the reputation of the industry was seriously damaged by naive and dishonest operators overselling their products or seeking to take advantage of generous tax credits and a gullible public.[1] These problems left a legacy of doubt and skepticism toward wind power that is only now beginning to fade.

Not surprisingly, the late eighties marked a period of painful readjustment for the wind industry, as tax credits were eliminated and fossil-fuel prices plunged, and many U.S. manufacturers of wind machines went out of business. Reflecting these changes, the amount of wind power capacity installed annually in California reached a peak of 398 MW in

1985, only to decline to 165 MW in 1991; just 5 MW are projected to be installed in 1992 (Gipe 1991, 1992).

Yet despite these reversals, wind power appears on the verge of a second and more enduring wave of development, spurred this time by a mature technology that promises to be less expensive than other sources of electricity, fossil and renewable. Although the California market has slowed considerably, several new utility-scale projects are under consideration in states such as Iowa, Maine, and Minnesota. Other countries, including the United Kingdom, Canada, Denmark, and Germany, are also pursuing numerous wind projects. Manufacturers of small wind turbines have already begun to see substantial growth in sales, mainly to developing countries, where these machines provide a reliable, low-cost alternative to diesel generators for supplying village power.

Over the long term, wind energy has the potential to supply a large fraction — perhaps 20 percent or more — of U.S. electricity demand. Aside from cost and market considerations, one of the obstacles to achieving this level of generation will most likely involve conflicts over available land. The lack of resource data on promising sites and the inherent variability of wind power will also be important considerations for electric utilities seeking to integrate this renewable resource into their existing power grids.

The Wind Resource

Winds are generated by uneven solar heating of the earth's surface. Because topographic features such as plains, hills, mountain passes, and large bodies of water can influence both wind direction and speed, winds can vary widely from one location to the next. In the middle of Ohio, for example, winds are fairly calm, but a few miles north near the shore of Lake Erie, they can be strong enough to knock a person down. The windiest areas in the United States tend to be found along the east and west coasts, along ridge lines in the Rocky and Appalachian mountain systems, and in a wide belt stretching through the Great Plains.

The amount of wind energy theoretically available for use in the United States is enormous: According to one estimate, about 3,000 EJ, or 40 times current annual energy consumption, are dissipated in winds annually (NSF 1972). Of course, only a small fraction of this resource could be exploited because of physical constraints on available land and

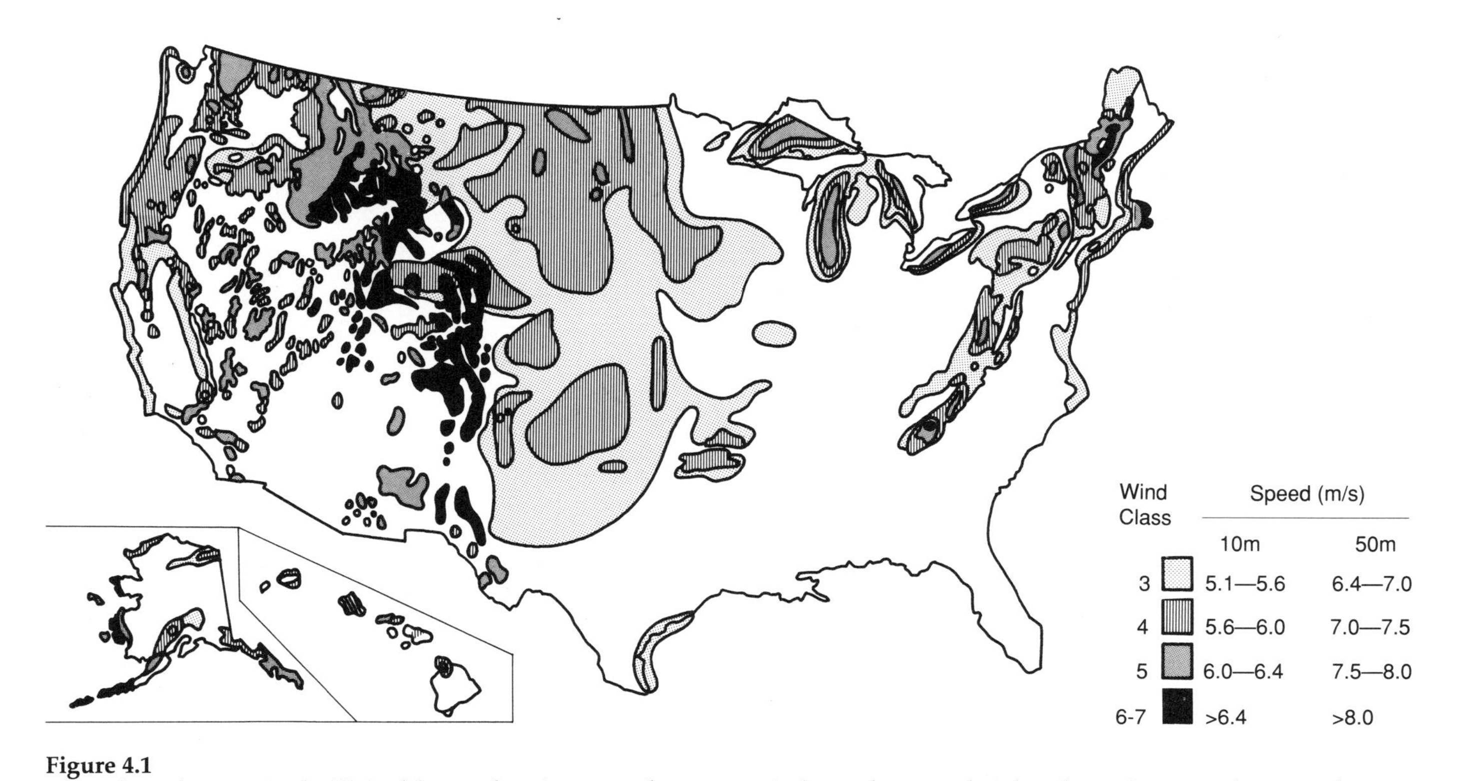

Figure 4.1
Map of windy areas in the United States showing annual average wind speeds at two heights above the ground — 10 m (33 ft) and 50 m (164 ft). Class 5, 6, and 7 areas are generally suitable for development with today's technology, whereas class 3 and 4 areas should become available with improvements in technology. Source: Pacific Northwest Laboratory.

the efficiency of energy extraction. But while estimates of the *recoverable* resource vary widely, the consensus is that it is very large. One review of wind studies in the seventies concluded that between 10 percent and 40 percent of U.S. electricity demand could realistically be supplied by wind power on land not currently used for other purposes (Kendall and Nadis 1980). As we will see, more recent work has considerably refined this estimate, although the broad conclusion remains unchanged.

Cost will play a central role in determining how much wind power utlimately can contribute to U.S. energy supply, and the cost of wind power is strongly dependent on the characteristics of the local wind resource. Since wind energy production is proportional to the cube of the wind speed, a 10 percent increase in speed yields roughly a 30 percent increase in the output of a wind turbine. At the same time, not all windy sites are suitable for wind power development. Some are on steep and rocky terrain, others are in scenic forests, still others are in densely populated coastal areas. Thus, a fair estimate of the practical U.S. wind speeds resource must take into account both the geographic distribution of winds and possible siting constraints.

The most detailed analysis of this issue so far was published in 1991 by researchers at Pacific Northwest Laboratory (Elliott, Wendell, and Gower 1991). In this study, the authors first estimated the amount of land area in the 48 contiguous states supporting wind speeds in each of seven classes, using a national wind resource data base (PNL 1986). Then they grouped areas into five categories of land use, environmental (e.g., national parks), urban, forest, cropland, and rangeland. Finally they constructed four scenarios that took into account varying degrees and types of possible land-use exclusions and thereby arrived at a range of estimates of the practical wind resource in each class.[2]

The authors found that the windiest areas of the United States (class 5 and above) could support enough wind power capacity to provide 18 to 53 percent of the electricity consumed in 1990. The lower figure represents the most severe assumptions of land-use exclusion (that is, all land excluded from development except 90 percent of rangeland), and the upper figure represents no exclusions at all. This resource is highly concentrated in a few Great Plains states (notably Montana, North Dakota, and Wyoming), however, raising the thorny question of how to deliver the power to other parts of the country where it is more needed.

In contrast, class 3 and 4 areas are distributed much more widely around the country and could support even greater wind power capacity than the class 5 to 7 areas. The study found that the total wind generation potential in these areas ranges from 1.7 to 6 times current U.S. electricity demand. Only class 5 to 7 areas are economical for wind power development today (barring the inclusion of environmental costs in utility planning), but class 3 and 4 areas may soon become available if advanced wind turbines under development meet expectations.

The feasibility of developing a particular wind site depends not only on the average wind speed and current land use, but also on more complicated factors such as daily and seasonal variations in wind speed and proximity to existing transmission lines. The value of electricity generated by a wind farm (a collection of wind machines) may be reduced if winds are out of synch with utility loads or increased if the opposite is true. These considerations are reflected in what is known as the capacity value, or load-carrying capability, of wind power, which is the amount of conventional generating capacity displaced per unit of wind capacity (discussed later). Similarly, wind power will be considerably more expensive if the site chosen for development is located far from existing transmission lines; conversely, at some sites, wind turbines may actually reduce loads on a utility's transmission and distribution system, thus resulting in a significant credit. Unfortunately, there has been relatively little analysis of issues such as these, and their effect on estimates of the practical wind resource is consequently difficult to assess.

Wind Technology

Wind turbines are deceptively simple machines consisting of blades, rotor, transmission, electrical generator, and control system, all mounted on a tower. There is presently a market for two size ranges, small (0.25-50 kW) and intermediate (50-500 kW). Several experimental machines of multimegawatt size were built during the height of the Department of Energy's wind research program in the late seventies and early eighties, but because of their high unit cost and history of technical problems they have not been developed for the commercial market, although research on them continues in Europe (Dodge 1992).

Intermediate-Size Wind Turbines

Intermediate-size wind turbines are designed for the bulk electricity market. Deployed in vast wind farms containing thousands of machines on towers 30 to 50 meters tall, they are an impressive sight to all who

Figure 4.2
Large arrays of wind turbines like these generate 1 percent of California's electricity, or enough to meet the residential needs of about 600,000 people. Source: US Windpower.

behold them. Modern intermediate-size wind machines have a peak capacity of 50 to 200 kW, but the trend is toward larger machines, and the optimum economic size is probably in the 300 to 750 kW range (McGowan 1991). Most intermediate-size machines now in operation have a two- or three-bladed rotor mounted on a horizontal axis that turns at a constant rotor speed, allowing for constant-frequency power to be supplied to the grid. In some horizontal-axis machines, the rotor is downwind of the tower and free to align itself with the wind, but in most the rotor alignment, or yaw, is actively controlled to keep it upwind of the tower or to avoid the tower's turbulent wake. A small number of wind turbines are of the eggbeater-shaped Darrieus design, in which curved blades are mounted on a vertical axis. This design has the unique feature that it is entirely insensitive to wind direction.

Although wind turbines appear simple at first glance, there are many subtleties to their design that have proved the downfall of more than a few companies. Reliability was initially a serious problem as components proved unexpectedly susceptible to fatigue, vibration failure, and mechanical breakdowns. Instead of the expected 20-year lifespans, some turbines lasted only 5 years, while operating and maintenance costs were also higher than expected, sometimes up to 4¢/kWh (Smith and Ilyin 1991).

Incremental improvements over the years have greatly increased the reliability and efficiency of wind machines, however. With more rugged designs, better choice of materials, and more careful attention to maintenance, average wind-turbine availability has risen above 90 percent — a level typical of conventional power plants — and mature systems are now operating 95 to 98 percent of the time. And in 1990, California wind turbines achieved an average capacity factor (actual generation divided

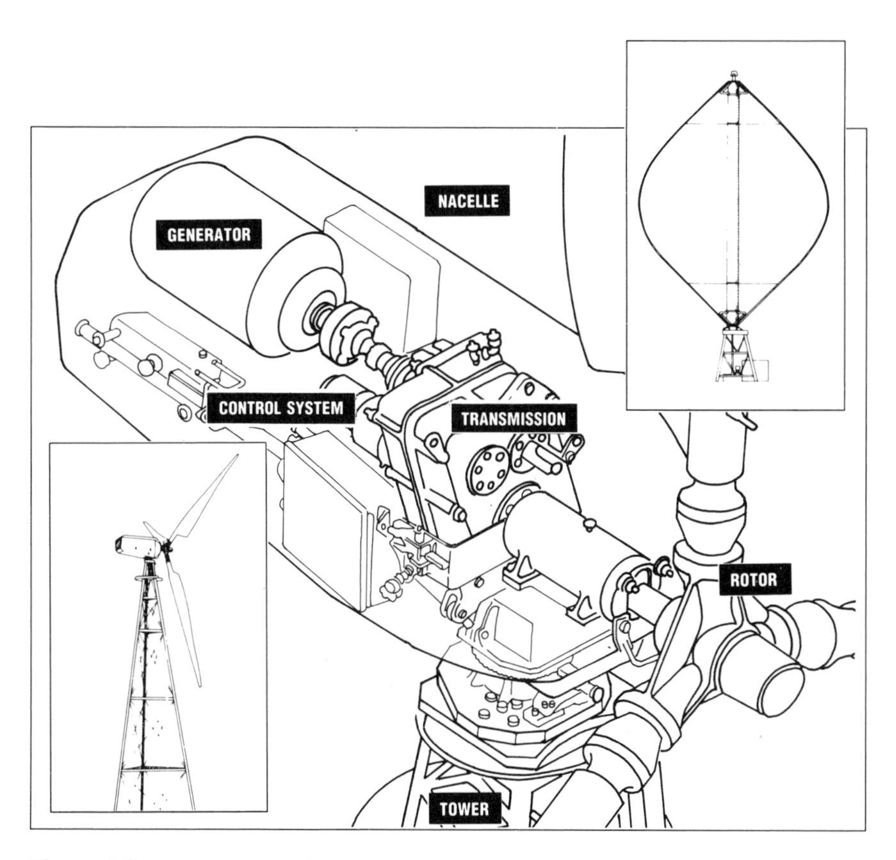

Figure 4.3
Typical wind turbine designs. In most wind turbines, the rotor is mounted on a horizontal axis (above and inset above left), but in some (known as Darrieus machines) it is on a vertical axis (inset above right).

by maximum generation at peak capacity) of 20 percent, up from 13 percent just four years earlier, while turbines installed since 1985 achieved an average capacity factor of 23 percent (Rashkin 1991).

As designs and performance have improved, the cost of wind power has declined dramatically. As of 1990, the levelized cost of electricity generated by wind turbines in operation in California was 5-8¢/kWh, depending on the wind site, down from more than 25¢/kWh in 1981. During the same period, the average price per installed kilowatt of intermediate-size wind turbines declined from $1,300-$2,000 to $950-$1,100 (Gipe 1991). Operations and maintenance costs have also been brought under control, and now average about 1.4¢/kWh (EPRI 1990a).

Further efficiency improvements and cost reductions are possible and, indeed, necessary if wind power is to become more widely successful around the country. Progress is being made continually as more advanced wind turbine designs move from development into production. One such design, slated for initial production in 1993, is being developed by an unusual consortium that includes US Windpower, the leading U.S. wind turbine manufacturer, the Electric Power Research Institute (EPRI), and several utility companies led by Pacific Gas and Electric. In addition to employing advanced airfoil shapes for its blades and improved power electronics and controls, this machine uses a variable-speed generator to produce electricity more efficiently over a wider range of wind speeds than is possible with fixed-speed generators. A power converter will convert variable-frequency electricity from the generator into the constant-frequency electricity required by utilities.[3] Besides improving efficiency, the use of variable speed generation should increase the lifespan of the machine by decreasing mechanical stresses on the rotor and drivetrain.

As of the end of 1991, a prototype of this advanced wind turbine, which will be rated at either 300 kW or 400 kW, had been installed in Altamont Pass, and two more were expected to be installed somewhere on the eastern side of Lake Ontario in 1992. Once in full production, this machine is predicted to cost $735/kW (at the 400 kW rating) and generate electricity at a levelized price of under 5¢/kWh in a 7 m/s (16 mph) average wind (EPRI 1990a, Ellis 1991).[4]

The declining cost of wind power does not mean it will have an easy time penetrating utility markets, however. Initial wind power development in California took advantage of long-term, fixed-price contracts (Interim Standard Offer 4) offered by electric utilities in the early eighties.

Figure 4.4
Small wind turbines of 1 to 10 kW capacity are ideal for generating electricity in rural or underdeveloped areas without an electricity grid. Requiring less maintenance than their larger cousins, these machines are becoming a preferred option over diesel power plants for third world villages undergoing electrification. Source: Bergey Windpower.

All of these contracts have run out, so new bids from wind power in California — as in the rest of the country — must compete against utilities' avoided costs, which are currently quite low because of excess capacity and the low price of natural gas. Wind power's competitive position is improving all the time, however, and utility markets are beginning to open, largely on the strength of US Windpower's new turbine. A few recent developments include:

• Iowa-Illinois Gas and Electric Co. and US Windpower have agreed to enter a joint venture to develop the first major wind energy project in Iowa or a nearby state, with an expected initial capacity of 250 MW.

• The Sacramento Municipal Utility District (SMUD) has accepted a bid by US Windpower to build a 50 MW wind farm using the new, variable-speed turbine.

• Northern States Power of Minnesota plans to build a 10 MW wind farm along Buffalo Ridge in the southwestern part of the state to gain experience with this technology.

• US Windpower is considering building a 50 MW wind farm in northern Maine.

Over the long term, the future of utility-scale wind generation seems very bright indeed, as long as siting conflicts (discussed later) do not prove too severe. Ultimately the cost of energy at sites with hub-height wind speeds of 6.8 to 8.5 m/s (15 to 19 mph) is expected to fall to 3-5¢/kWh, well below the price of most conventional alternatives (Hock, Thresher, and Cohen 1990).

Small Wind Turbines

Many of the same developments that have brightened the prospects of intermediate-size wind turbines have affected small wind turbines as well. Materials and designs have improved, while at the same time manufacturers and operators have gained valuable experience with these machines under a variety of conditions. Perhaps surprisingly, small wind turbines require less maintenance than their larger cousins, mainly because of their simpler design. In most machines, the rotor drives the generator directly — there is no intervening transmission and gears as in larger machines — so that mechanical wear is greatly reduced and there is no need for oil changes. For some, regular maintenance is required only every four or five years, when tape applied to the leading edges of the blades must be replaced (at a cost of around $50). Occasional lightning strikes may damage a generator, but overall operating and maintenance costs are very low — typically around 0.3¢/kWh (Bergey 1991).

Small wind turbines are intended not for the bulk electricity market but for small-scale, residential and commercial, grid-connected and remote applications. Current installed costs for 0.25 kW to 10 kW systems range from $1,370 to $3,580 per kilowatt, but in small-scale remote applications where winds average above 9 mph they are often the least expensive electricity-supply option. They are ideally suited to the rapidly growing market for village power in developing countries, where they compete with photovoltaic systems, diesel generators, and grid extension. For rural homeowners and farmers who want to reduce their electricity bills, small wind turbines can be an attractive option. Since in this case they are displacing electricity at retail prices, the economics can be as favorable as for the larger systems.

In the United States the market for small wind turbines is likely to be rather limited, probably less than 5 MW of annual sales, unless new tax credits or low-interest loan programs are reinstated to make the initial capital expenditure less burdensome. One possible application that has not yet been developed is the reduction of peak loads in parts of the utility network where transmission and distribution systems are aging or approaching their load limit. This is identical to the photovoltaic application being explored by Pacific Gas and Electric (discussed in the previous chapter). Hybrid wind/photovoltaic systems might well prove better suited to this application than either type of system alone.

Intermittency

Conventional wisdom holds that the intermittent nature of wind power will relegate it to an incidental role in the nation's electricity supply, at least until reliable and low-cost energy storage becomes widely available. This assessment is misleading, however. The many complex and subtle issues raised by wind power's variability are not widely understood. Indeed, one of the problems currently facing wind power developers is the lack of appreciation even among electric utilities of wind power's full value to an electric utility system.

In the first place, although wind speed and direction fluctuate widely, they generally follow daily and seasonal patterns that are surprisingly predictable (as sailors following trade winds discovered long ago). Using statistical methods, it is possible to determine from several years of wind speed measurements the amount of power a wind farm can be expected to deliver at any given time of day or year. The more important issue for utilities is not the predictability of wind power but the degree of correlation between the wind power output and utility loads. In areas where there is a good match of wind and load — that is, winds peak at the same time utility loads are highest — then a wind turbine is said to have a high capacity value, which is defined as the amount of "ideal" generating capacity displaced per unit of "real" capacity added. If wind and load are out of step, however, then the capacity value may be very low, and wind power will displace only fuel. Of course, the actual capacity value can — and usually does — fall somewhere in between these two extremes (Flaim and Hock 1984).

These principles are illustrated by the experience of Pacific Gas and Electric, a summer-peaking utility whose loads are dominated by air conditioning and irrigation-water pumping. Winds in the Altamont Pass region where many wind machines are located provide a good *seasonal* match to the utility's loads, with about 80 percent of the total wind generation coming between May 1 and October 31. On a daily basis, however, Altamont winds do not hit their peak until 9 or 10 PM, too late to meet PG&E's peak power needs. Thus Altamont Pass turbines have a capacity value of only 10-40 percent, that is, one kilowatt of wind capacity will displace only about 0.1-0.4 kW of conventional peak capacity. In Solano County, just east of San Francisco, however, winds peak around 4 PM, coinciding almost exactly with PG&E's needs and thereby giving these turbines an extremely high capacity value of around 80 percent (Smith 1990). Unfortunately, even when utilities recognize and accept that wind can displace conventional capacity, most do not have access to sufficiently detailed information on local wind resources to predict accurately the potential load-carrying capability of wind power on their systems. The lack of site-specific resource data is consequently one of the most serious obstacles facing the wind-turbine industry in its efforts to expand to areas outside California.

If wind power is ever to contribute a very large fraction of U.S. electricity supply, then some form of storage or backup capacity will be needed to correct for the inevitable mismatches between wind power output and demand. Exactly how much storage or backup is needed will vary widely from one utility system to the next. In theory, the storage requirement can be greatly reduced by linking a number of widely separated wind farms into a single grid, so that a wind drop in one area would often be offset by an increase in another. Combining both wind and solar generation might also be advantageous. For example, peak solar insolation in California tends to precede peak electricity demand in the summer by a few hours, while Altamont wind generation tends to lag behind. The combination of wind and solar power is thus almost perfectly synchronized with PG&E's needs (Smith 1990).

In any event, storage will not be an issue in the near term as electric utilities are well equipped to handle variations in wind power output when it is only a small fraction of the total electricity supply. (After all, utility loads vary by as much as 50 percent throughout the day.) According to some studies, storage will not be needed until wind power constitutes at least a few percent, and perhaps as much as 20 percent (assuming an interconnected network of windfarms), of the total supply (Kendall and Nadis 1980, Grubb 1988). Utility systems such as those in

Public Attitudes toward Wind Power

Wind power is, in the most literal way, public power: The turbines generate electricity out in the open for all to see. Such visibility may engender in some a sense of responsibility for electric power use, but it also attracts attention and, from many, opposition.

In a study of public opposition to wind power projects in California, Thayer and Hansen (1991) identified four main issues that will likely face all future wind projects. The most crucial is competition with housing development. Both require large open areas, a dwindling resource in many states, and for home builders, rural or suburban character can be an important selling point, an image that a large wind farm may dispel. For example, residents opposing the Cordelia Hills project in Solano County, northeast of San Francisco, "simply did not want to see turbines sited on the land visible from their windows," even though the hills chosen for the project were "overgrazed, contained numerous electronic relays and transmission lines, and had no stands of native grass species." The visual impact of hundreds of wind turbines set atop tall towers along windy ridges, hilltops, or mountain passes was also a key factor sparking objections to wind development at Tejon Pass in Los Angeles County, considered to be one of the most scenic areas close to Los Angeles. The 458-turbine project was eventually defeated.

In keeping with the public nature of wind power, opponents often described abandoned turbines or what appeared to be nonworking turbines they had seen at existing wind farms as a factor in their opposition. This is what Gipe (1991) calls the "missing tooth" phenomenon, in which the eye is drawn not to the 99 turbines spinning away in the wind but to the 1 that sits idle.

Yet in appropriate settings wind power development has moved forward with little or no opposition. For example, US Windpower was successful in its bid to construct a 400-turbine wind farm at Montezuma Hills in Solano County, California, mainly because the site was located 15 miles from the nearest freeway and 10 miles from the nearest town.

the Pacific Northwest that have relatively large amounts of hydropower capacity (which can be used for pumped storage or load-following) can support even larger amounts of intermittent renewable generation with no additional storage. One study indicates, for example, that up to one-half of a hypothetical California utility's peak loads and one-third of its annual generation could be provided by solar and wind power with no need for additional storage (Weinberg and Williams 1990).

Environmental and Siting Issues

It is hard to imagine an energy source more benign to the environment than wind power, for it produces no air or water pollution, involves no toxic or hazardous substances (other than those commonly found in large machines), and poses no threat to public safety. And yet one of the most serious obstacles facing the wind industry has to do precisely with its environmental impacts, most importantly conflicts over the use of land.

The land required for wind generation is a confusing issue that must be carefully defined. Most studies assume that wind turbines will be spaced a certain distance apart and that all of the land in between should be regarded as occupied. This leads to some quite disturbing estimates of the land area required to produce substantial quantities of wind power. According to one widely circulated report from the seventies, for example, generating 20 percent of U.S. electricity demand from windy areas would require siting turbines on 18,000 square miles, an area about 7 percent the size of Texas (GE 1977). According to this study, wind turbines would generate up to 1.8 MW per square kilometer of land occupied.

In reality, the wind turbines themselves occupy only a small fraction of this land area, and the rest can often be used for other purposes or left in its natural state. For this reason, wind power development is ideally suited to farming areas. In Europe, farmers plant right up to the base of turbine towers, while in California cows can be seen peacefully grazing in their shadow. The leasing of land for wind turbines — far from interfering with farm operations — has brought substantial benefits to landowners in the form of increased income and land values. Perhaps the greatest potential for wind power development is consequently in the Great Plains, where vast stretches of farmland could support hundreds of thousands of wind turbines with little likelihood of significant protest.

In other settings, however, wind power development can create serious conflicts. In forested areas it may mean clearing trees and cutting roads, a prospect that is sure to generate controversy and public opposition, except possibly in areas where heavy logging has already occurred. Even in otherwise developed areas, wind projects often run into stiff opposition from local residents, some of whom may regard them as unsightly and noisy.

In California, bird deaths from electrocution or collisions with spinning rotors have emerged as a problem at Altamont Pass, where more than 33 threatened golden eagles and 75 other raptors such as red-tailed hawks died or were injured during a three-year period. Studies underway to determine the cause of these deaths and possible preventive measures could have an important impact on the wind industry. According to one report, "Wind energy's ability to emerge as a positive symbol of environmentally responsive energy production will be seriously jeopardized if wind turbines become popularly associated with the death of raptors" (Thayer and Hansen 1991).

Problems such as these are not unique to wind power. They increasingly affect every aspect of utility business, in which it has become notoriously difficult to site new power plants and transmission lines. But with imagination, careful planning, and — perhaps most important — direct, early contacts between the wind industry, environmental groups, and affected communities, there is no reason to believe the obstacles cannot be overcome.

Notes

1. For example, one company, KSI & Associates of Milan, Minnesota, marketed a small wind machine in 1981 called the Wind Mule, which they claimed could produce 6 kW of power in a 12 mile-per-hour wind. In fact, a machine of the Wind Mule's size could produce not 6 kW, but about 60 watts, a discrepancy of a factor of a hundred. The company, after accepting orders and selling distributorships, subsequently disappeared (Bergey 1990).

2. The wind resources cited in chapter 3 were taken from this study's "moderate exclusion" scenario.

3. Variable speed generators have been used since 1980 on small wind turbines.

4. If the hub height is 30 meters (typical for current intermediate-size machines), this wind speed corresponds to a class 5 site. If the hub height is raised to 50 meters, it corresponds to a class 4 site.

5 Biomass

For all but a fraction of recorded history, biomass has been humanity's predominant source of fuel. Burning wood, dried plants, plant oils, animal dung, and even animal fat met most heating, lighting, cooking, industry, and transportation needs until the latter part of the 19th century. Indeed, in much of the developing world, biomass — principally firewood — still supplies most heating and cooking needs for rural populations.

In the United States, biomass energy use steadily declined until the sixties, but since then it has expanded to equal about 4 percent of energy demand in 1990 (EIA 1991a). Most of this increase is due to the paper and pulp industry, which today satisfies more than half of its energy needs using mill wastes. The most important future role for biomass, however, will most likely be the production of liquid and gas fuels such as ethanol and methanol, which generate less pollution than coal and oil. There is already a small industry turning corn into ethanol for mixing with gasoline, and although this particular approach has serious drawbacks, the development of improved conversion processes and less expensive energy crops could open large new markets for such biofuels, particularly for transportation. Since they can be burned in only slightly modified boilers, combustion turbines, and automobile engines, they could provide a crucial bridge between today's fossil-fuel economy and one powered by truly clean sources such as solar hydrogen and electricity.

Before biofuels can come into their own, however, a number of complex environmental and socioeconomic issues must be addressed. These range from the health effects of air pollution and disposal of solid wastes generated by biomass combustion to the impacts of energy crop cultivation on soil and water quality. Not all of the impacts will necessarily be negative. Cultivating energy crops, for example, may prove to be a valuable source of income for farmers as well as a means for reducing soil erosion and water runoff on erodible land. Such issues can only be touched on in this chapter, but they will have a powerful bearing on the development of biomass as a long-term energy resource.

The Biomass Resource

As with most other renewable resources, the energy in biomass ultimately comes from the sun. Through the process of photosynthesis, nearly all plants, from the majestic redwoods to the diminutive algae, use the energy of sunlight to manufacture carbohydrates, which include sugars, starch, cellulose, and other compounds. Carbon dioxide plays a vital role in this process. As plants grow, they absorb the gas from the atmosphere and give off oxygen. When they die and decompose, much of the carbon stored in the plant material is transferred back to the atmosphere as methane or carbon dioxide. Through this cycle of growth and decay, plants play a vital role in regulating the natural greenhouse effect.

When most people think of energy from biomass, they imagine cutting down trees and burning the wood in fireplaces or cast-iron

Figure 5.1
Energy farms like this one in Minnesota, which grows aspen, could supply a substantial fraction of U.S. energy demand using idle or converted cropland. Source: University of Minnesota Extension Service.

stoves. In reality, there are many potential sources of biomass besides forests and many different ways of converting this biomass into energy besides direct combustion. The resource is usually divided into three categories: wastes, standing forests, and energy crops, the latter including trees and herbaceous plants (grasses and legumes). Calculating the energy potential of each source involves a number of estimations, including the amount of biomass that can be produced and collected at an economical price, its energy content, and, especially if converted to a liquid or gas fuel, the conversion efficiency. Uncertainties in all three lead to wide-ranging estimates of the total biomass resource. The estimates here are intended to give an idea of the general magnitude of the resource, not to set absolute limits.

Wastes

Vast quantities of organic wastes are produced by both cities and industries. Extracting energy from these wastes can accomplish a number of useful things. First, since wastes contain significant energy stores that would otherwise be discarded, making use of them increases the overall efficiency of the economy. Second, it can save money, as wastes are often more expensive to dispose of than to burn. Finally, it can extend the life of landfills, which are in increasingly short supply because of difficulties in opening new landfill sites.

The pulp and paper industry is the leading user of wastes for energy in the United States, in 1990 generating about 1.4 exajoules (EJ), or 1.6 percent of U.S. energy demand, from wood scraps, unusable trees, bark, and wood-pulp residue (API 1991). Not much waste goes unused at large pulp and paper mills, however, so little can be done to expand the use of this resource (although, as we will see, it is possible to increase electricity production at mills through the use of more efficient technology). A good deal of logging residues (tree stumps, branches, and leaves) are left in the field in forestry operations, however, and could be used for energy. In Minnesota, Wisconsin, and Michigan, for example, 18 to 32 percent of all the timber cut in 1988 was left in place (USDA 1991a, 1991b, 1991c). If the typical fraction for the United States as a whole is 25 percent, there may be 60 to 70 million metric tons of annual logging residues available, implying a potential resource of about 1.3 EJ per year.[1] Not all of this resource can be recovered, however, as some may be difficult or expensive to collect or should be left in place to help hold forest soil, replenish nutrients, and catch rainfall runoff.

Table 5.1
Consumption of biomass energy in 1987, in exajoules per year. Source: Klass (1988).

Wood and Wood Waste	**2.87**
Industry	1.95
Residential	0.89
Commercial	0.02
Utilities	0.01
Agricultural and Industrial Waste	**0.04**
Municipal Solid Waste	**0.12**
Landfill Methane	**0.01**
Ethanol	**0.07**
Other	**0.01**
Total	**3.12**

Agricultural wastes, such as plant stalks and leaves, are also a significant potential source of energy. Like logging residues, these tend to be widely dispersed, and it is usually desirable for ecological reasons to leave some or all of them in place. But at least a portion of some types of crop residues could be collected without significant impact on soil quality. For example, a study of the federal Western Regional Biomass Energy Area (including 13 states west of the Mississippi but excluding the Pacific Northwest) found that about 16 percent of the available crop residues, or 22 million metric dry tons, could be collected and used (Tyson 1991). Other agricultural wastes are produced by the food processing industry. For example, the crushed plant residue that remains after sugarcane processing, called bagasse, supplies about 9 percent of Hawaii's electricity (Takahashi et al. 1990). Overall, the energy available from agricultural wastes is estimated to be about 1 EJ per year (DOE 1988b, IEA 1987).

Crop surpluses are another form of waste, this one fostered by government subsidies that encourage farmers to overplant their fields. Grain surpluses, principally corn, are the main source of feedstock for the ethanol industry, whose product is sold in a gasoline-ethanol mixture known as gasohol. Theoretically, corn and other grains could supply around 3 billion gallons of ethanol annually, or enough to displace about

3 percent of U.S. gasoline consumption, before causing grain prices to rise substantially (Lynd et al. 1991). Since so much energy is used to make both ethanol and corn, however, this is probably not a sensible option for the United States. (The net energy balance of biofuels is discussed further below.)

Another source of biomass is municipal solid waste, or in plain words, city trash. In 1986, American cities and towns generated 144 million metric tons (158 million short tons) of trash, and this production is growing at a rate of about 2 percent per year (Bureau of Census 1990). Of this waste, about 60 percent consists of organic materials such as paper, yard wastes, and wood, implying a potential energy resource of around 1.7 EJ per year. Some of this is already being used for energy. In California, for example, 250,000 tons of urban wood waste — including landscape brush, demolition materials, pallets and plywood — were consumed by small power plants in 1990 (Miles and Miles 1991). Throughout the United States, however, only about 10 percent of municipal solid waste is converted to energy; another 10 percent is recycled and the rest is disposed of in landfills (OTA 1989). Once buried in a landfill the waste may still produce usable energy in the form of methane, which can be captured and burned to produce electricity. Potential energy production from this and other resources such as sewage and industrial solid wastes is estimated to be about 0.3 EJ per year (Klass 1988).

Standing Forests and Energy Farms

Although biomass wastes are an important energy resource, much more energy could be obtained by exploiting America's rich resources of land for growing energy crops. Of course, the firewood we use in our homes today comes from natural energy "farms" — forests. But natural forests are relatively inefficient energy producers. One hectare of commercial forest (i.e., forest capable of supporting commercial rates of wood production) produces on average about 2.5 tons of dry wood per year,[2] implying that harvesting 40 million hectares (100 million acres, or about one-fourth of current U.S. forestland) would yield no more than about 2 EJ per year. Although this yield could undoubtedly be increased through more efficient and intensive management, competition from the forest-products industry will tend to limit the amount of wood diverted to energy use from existing commercial forests, and environmental concerns about clear cutting and the destruction of old-growth forests could prevent the opening of much new forestland to production.

For these reasons, attention is being focused on developing fast-growing varieties of trees and plants for cultivation on currently unforested land. Research programs directed by the Department of Energy at Oak Ridge National Laboratory are aimed at identifying suitable woody and herbaceous plants for use as energy crops and developing efficient methods for growing and harvesting them. Among potential tree species, hardwoods such as hybrid poplars, eucalyptus, silver maple, and black locust appear most promising. Depending on the species and location, the trees would be harvested every two to eight years; hence their name, short-rotation woody crops (SRWC). Since they are coppicing species (new shoots grow from cut stumps), planting is only required once every three or four harvests. At present achievable yields of about 5 to 15 dry tons per hectare per year (t/ha/yr), the Department of Energy estimates the cost of energy from SRWCs (including planting, harvesting, and transportation) to be about $3.25/GJ. Ultimately researchers hope to double or triple this yield in commercial plantations, to perhaps 25 t/ha/yr, bringing the cost down to $2/GJ, roughly the current price of coal (Bull 1991, DOE 1990a).

Several types of herbaceous plant are also being considered as energy crops. Research is focusing on fast-growing annual and perennial grasses — sorghum, switchgrass, Napier grass, and others — that can be harvested with existing farm equipment. These would be grown much like ordinary food crops, but the perennial varieties offer significant advantages in that their roots hold erodible soil and, once established, they are highly resistant to weeds. The Department of Energy supports research in several regions of the country, testing energy crops under a variety of conditions of climate, soil type, and fertilizer use. Herbaceous crop yields are somewhat higher than SRWC yields, commonly reaching 12 to 40 dry tons per hectare per year.

Assuming that the market encourages farmers to grow energy crops — a critical question — the amount of energy that can be produced will ultimately be limited by the amount of available land. Even using advanced cultivation and conversion methods, producing enough biofuel to meet all U.S. transportation needs — currently 23 EJ per year — would mean converting around 80 million hectares, an area equal to about two-thirds of harvested cropland, to energy production.[3] Yet the potential contribution of energy crops is nonetheless important, for there is a large amount of land available that is not now under cultivation. Some 40 million hectares of farmland lie idle in any given year, some as part of the

Soil Conservation, Conservation Reserve, and other federal programs. In addition, up to 60 million hectares of pastureland, rangeland, and forestland have been classified by the U.S. Department of Agriculture as having "medium to high" potential for conversion to cropland. According to one estimate, if these areas were converted entirely to energy production, they could provide anywhere from 20 EJ to 56 EJ annually, depending on average crop yields (Cook, Beyea, and Keeler 1991).

Of course, if there were a market for energy crops, farmers might very well choose to grow them on prime land to achieve the highest possible yields and profits. This could result in some displacement of conventional food production. This outcome was suggested by a recent study of the economics of large-scale energy crop cultivation, which found that

Table 5.2
Estimates of potential energy production from biomass wastes, forestry, and energy crops. The realistic potential is probably considerably smaller than the upper bounds suggest.

	Available Land Area (Millions of Hectares)	Energy Production (EJ/Year)	Fraction of 1990 U.S. Energy Demand
Energy Crops[a]	**101**	**20-56**	**22-63%**
Cropland	40	8-22	
Pastureland	24	5-13	
Rangeland	20	4-11	
Forestland	17	3-10	
Forestry and Wastes[b]		**15.4**	**17%**
Wood and Wood Wastes		11.0	
Crop Residues		1.1	
Aquatic Biomass		0.8	
Municipal Solid Waste		1.9	
Landfill Methane		0.2	
Industrial Solid Waste		0.2	
Other Waste		0.2	
Total		**35-71**	**39-80%**

[a] From Cook, Beyea, and Keeler (1991), based on estimates of idle cropland and convertible pastureland, rangeland, and forestland areas, and assuming average biomass yields of 10-28 dry tons/hectare/year, or 200-560 GJ/ha/yr based on an energy content of 20 GJ/dry ton.
[b] From Klass (1988).

producing 7.6 EJ of methanol per year nationwide by 2030 would require planting energy crops on 31 million hectares. Although about 19 million hectares of this would consist of idle cropland, pastureland, and rangeland, about 15 million hectares of harvested cropland would be used as well, resulting in a modest decrease in food production and a corresponding increase in food prices. Overall, there would be major economic benefits for farmers, who would earn $3 billion to $11 billion more annually from combined sales of energy and food crops than they would otherwise earn from food crops alone (Tyson 1990).[4]

The study illustrates both the significant potential of energy crops and the perils of ignoring ecological considerations. One of its findings was that, under current projections of energy crop yields and production costs, farmers would most often choose to cultivate annual species such as sorghum, which are grown in a manner similar to corn. But although sorghum may prove to be the best energy crop from a profit standpoint, its cultivation on a large scale would greatly increase soil erosion. Perennial species and short-rotation trees do not have this drawback, but their production costs would probably be higher and consequently government regulations or incentives might be needed to encourage their cultivation.

Woody and herbaceous energy crops would most likely be converted to methanol or ethanol for use in automobiles, but other crops could be raised for their oils, fatty acids and other complex molecules that can be substituted for diesel and fuel oil. One crop that could yield large quantities of fuel oil at a competitive price — because it can be grown in the off-season in southern climates — is winter rapeseed. Aquatic plants and microalgae grown offshore, in marshes, or in desert ponds fed by saline groundwater could also produce a variety of fuel oil substitutes, although their development as a practical fuel source is regarded as a more distant prospect. The Department of Energy estimates that up to 2 EJ could be derived from both sources (DOE 1988b).

While it is possible to assign numbers to potential energy crop resources, it is far from clear how a market for them can be created. Research can bring the cost of biomass production down, but at the moment there is no industry invested in developing energy crops as there are industries invested in, say, wind power and photovoltaics. Moreover, if energy crops are to succeed, farmers must become fuel suppliers catering to an entirely different set of customers (e.g., electric utilities) than that with which they are accustomed to dealing. This transformation is unlikely to happen on its own without encouragement

Figure 5.2
The paper and pulp industry is the largest consumer of biomass, using black liquor, bark, and other process wastes to generate steam and electricity. This cogeneration plant is located in Springfield, New Hampshire. Source: Hamphill Power and Light.

from federal or state government. One obvious step is for states, with federal assistance, to set up energy farm demonstrations in conjunction with biofuel production facilities. At least one state, Minnesota, is proposing such a project, which could be either a 1,200 hectare (3,000 acre) hybrid poplar farm to supply an ethanol production facility, or a 20,000 hectare (50,000 acre) hybrid poplar farm to supply a 100 MW whole-tree burning power plant (Helgeson 1992).

Direct Combustion

Burning biomass directly is a crude but effective way to convert the chemical energy stored in plants into heat and, if desired, electricity. Direct combustion technology is well established. In 1990, home wood burning and the combustion of wood wastes by the forest-products industry together accounted for over 90 percent of the 3.3 EJ generated by direct combustion (EIA 1991b). Most industry wastes consist of black liquor, the residue left over after chemically pulping wood for paper production, while the rest is made up largely of bark and chipped wood. The pulp and paper industry is a leader in cogeneration — the simultaneous production of heat and electricity for industrial use — and in 1990 met about half of its own electricity needs in this fashion (API 1991).

In 1987, the latest year for which statistics are available, about 5.6 percent of American households burned wood as their main source of heat, while another 19 percent used wood at least occasionally, generating about 0.90 EJ in all (EIA 1991b). These statistics are somewhat misleading, however, since the open fireplaces used in many homes are highly inefficient. Modern, enclosed fireplaces, which draw the air directly from the outside and circulate it through a thermal mass before sending it up the chimney, perform much better, as do wood-burning stoves. In areas where wood is abundant and cheap, some institutions are realizing significant savings by switching to it. Three public schools in Vermont, for example, cut their yearly energy costs 55 to 80 percent just by converting from oil or electric heat to wood (Etkind, Hudson, and Slote 1991).

The production of electricity from biomass has greatly increased in the past decade. In 1988, about 5,100 MW of mostly wood-fired electric capacity (including cogeneration) were in operation, up from 250 MW in 1981 (Klass 1988). The vast majority of the increase is due to industrial cogeneration; electric utilities have shown only slight interest in wood-fired power plants, at least in part because the relatively small size of the plants — usually under 50 MW — and the need for specialized fuel-handling equipment results in relatively high capital costs ranging from $1,500 to $2,500 per kilowatt (CEC 1987).[5] In addition, conventional wood-fired power plants can be troublesome to operate because of problems such as alkali slagging, in which alkaline compounds such as potassium and sodium in the biomass become molten and deposit themselves on boiler surfaces.

More advanced concepts have been proposed that could make electricity generated from biomass considerably less expensive than the 5-7¢/kWh typical of today's technology. For example, burning whole trees instead of wood chips could permit larger plants to be constructed while eliminating the need for screens, hammer mills, chipping equipment, and other wood-handling equipment (EPS 1990). Several utilities have expressed interest in whole-tree power plants, and funding to operate a small prototype facility is being provided by the Electric Power Research Institute. Gasifying wood or other biomass and sending the gas through a combustion turbine is another promising approach (discussed later), as is co-firing wood or other biomass with coal in existing coal-fired power plants (DOE 1991a).

Municipal solid waste (MSW) is also consumed for energy, most often because cities do not know what else to do with it. The cost of trash disposal has been rising since the seventies as it has become more difficult to site new landfills and as existing landfills have reached capacity and closed. Almost 140 waste-to-energy facilities were in operation in 1990 (Kiser 1991). Most of these were mass-burn facilities, which consume raw waste, but some used what is called refuse-derived fuel (RDF), essentially chopped-up or pelletized waste with most of the metal, glass and other inorganic matter removed. RDF offers the advantage that it can be burned in conventional boilers and contains fewer toxic materials that can enter the air. RDF can also be burned with wood chips, coal, or both, in co-fired facilities.

Political and economic pressures to convert municipal waste to energy are sure to grow as waste-disposal costs continue to rise. It is unclear, however, how many waste-to-energy plants can be built in the future. Several that had been proposed were blocked by local opposition sparked by such concerns as air pollution and toxic ash disposal, the social stigma attached to living near a waste facility, and safety risks from increased truck traffic. As of 1988, 25 completed facilities had been permanently shut down for a variety of reasons, including environmental concerns and poor performance (Klass 1988). Many cities are turning to recycling as a way of avoiding, or at least delaying, the need for new incineration facilities, and indeed this approach may offer important economic and environmental advantages (EDF 1985, Newsday 1989).

Thermochemical Conversion: Methanol and Syngas

Heating biomass in an oxygen-deficient or oxygen-free atmosphere breaks down complex organic molecules into simpler solids, liquids, and gases. This method of fuel conversion has been known for millennia, since the first primitive people made charcoal from wood. Three basic techniques — gasification, liquefaction, and pyrolysis — have been developed to produce different types of fuels from carbon-rich biomass. Advances over the past decade have made it possible to manufacture a wide variety of substitutes for petroleum, natural gas, and petrochemical feedstocks with fewer problems than were experienced previously. In combination with advanced combustion technologies such as gas turbines, these biofuels could find wide application in the future, with fewer environmental impacts than are associated with direct combustion.

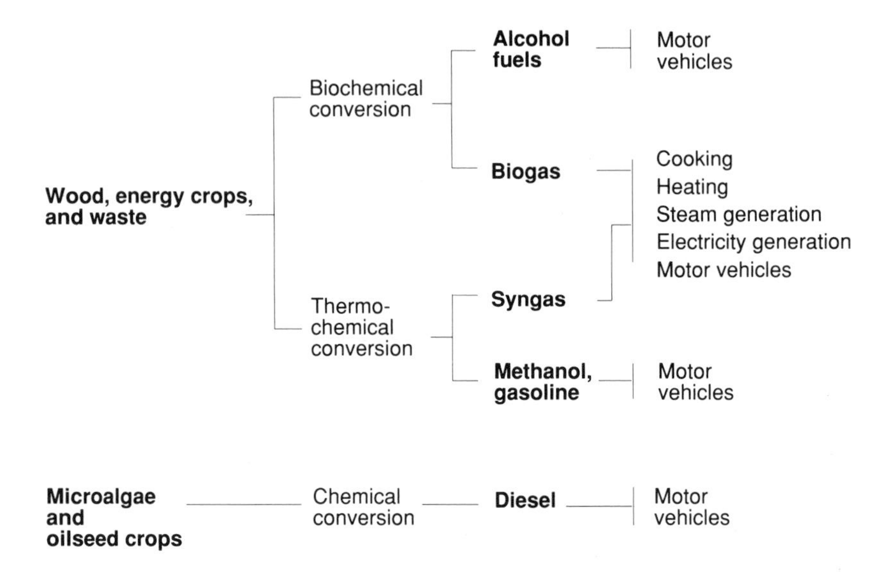

Figure 5.3
Pathways for biofuels production.

Gasification

Of the thermochemical processes, gasification is the most familiar and most widely used in the United States. The basic process has been known since 1792, when a Scottish engineer first gasified coal to light his home. Coal gas, also known as town gas, was distributed throughout Europe and the United States in the latter half of the 18th and early part of the 19th centuries for street lighting and home heating. This low-energy gas, now called syngas, is made by passing steam over a heated carbon source such as coal or biomass to produce a mixture of carbon monoxide and hydrogen gas. Using air in the process produces what is known as low-Btu gas (LBG), the most common type, with up to one-quarter the volumetric energy content of natural gas. Using pure oxygen yields medium-Btu gas (MBG) with up to one-half the volumetric energy content of natural gas. (High-Btu gas equivalent to natural gas can also be made, but the process is energy-intensive and far from economic.)

A modest gasification industry was established in the early eighties to supply the forest-products industry, which uses the gas generated by wastes to fire wood kilns and provide space heating. Although few gasification systems have been installed since fossil-fuel prices began falling in 1985, the potential market for syngas is nevertheless quite large, since existing gas- and oil-fired boilers can be readily adapted to burn it. Department of Energy research is directed towards maximizing the efficiency of gasification, and researchers hope to reduce the production cost from the historical $8/GJ down to $3.50/GJ — close to today's price of natural gas for industrial users — by the mid-nineties.

Biomass gasification may be ideally suited to integration with high-efficiency gas turbines for generating electricity.[6] A similar concept has been under investigation as part of the Department of Energy's "clean coal" program. A100 MW coal-fired integrated gasifier/combined cycle (IGCC) system has been operating at Cool Water, California, and a 167 MW IGCC system has been operating at Plaquemine, Louisiana. A major advantage of adapting this concept to biomass is that biomass gasification/generating systems could be less expensive and more efficient than conventional wood-fired steam-electric power plants. According to one estimate, a typical 50 MW system using a steam-injected gas turbine would cost about $1,000/kW and generate electricity at 40 percent efficiency (excluding losses in gasification), about half the cost and nearly twice the efficiency of conventional biomass power plants (Larson 1990).

Both the gasifier and gas turbine technologies are commercially available, but an integrated system has not yet been tested, for unlike clean coal technologies, "clean biomass" technologies have received only modest federal support.[7] One challenge is to ensure that the gas produced in the gasification process is sufficiently free of tars, oils, alkali compounds, particulates, and nitrogen oxides that can affect turbine performance or result in air pollution; another is to develop gasification systems that can handle larger volumes and a wider range of feedstocks (for example, wood chips and bagasse) than existing systems (DOE 1991a).

The sugarcane and pulp and paper industries present attractive opportunities for the biomass gasification/gas turbine combination. Analysis suggests, for example, that a 50 MW system operating year-round could generate all the electricity and steam needs of a sugarcane factory plus a large amount for sale to a utility. The electricity revenues for the factory (at a sale price of 5¢/kWh) would be about double the revenues from sugar sales at 1986 prices (Ogden, Williams, and Fulmer 1991). The same technology could be applied to wood, wood wastes, energy crops and municipal solid wastes, although efficiencies would vary depending on the type of feedstock. According to one assessment, if all U.S. pulp and paper mills were converted to advanced gasifier/generator systems (with black liquor as the fuel), the electricity produced would equal about 4 percent of U.S. electricity demand (Larson 1990).

A key problem in getting industries to adopt such technology is that factory and mill operators do not see themselves as being in the electricity-selling business. Although many practice cogeneration, it is largely to meet their own power needs, not to sell to utilities. Utility managers, on the other hand, do not see themselves as being in the cogeneration business and, moreover, are concerned that cogenerators may not be reliable electricity suppliers. This problem illustrates how divisions between utilities and industry can result in a failure to take advantage of cost effective options for generating electricity.

Pyrolysis

Pyrolysis is a thermochemical process that produces a mixture of solids (char), liquids (oils and methanol), and gases (methane, carbon monoxide, and carbon dioxide), with the proportions determined by the operating temperature, pressure, oxygen content, and other conditions. Although the technology is well established, commercially available products have tended to be costly and somewhat troublesome to use, since residual tars can foul gas systems, and the oils produced can be acidic and difficult to store. As a result, few companies have had commercial success with pyrolysis systems.

This picture appears to be changing as a result of advances in research. One promising application for pyrolysis that has emerged in the past several years is the production of petrochemical feedstock substitutes, such as phenol-based adhesives used in making plywood and particle board. "Fast" pyrolysis of sawdust for a few seconds at temperatures of 450°C to 600°C (840°F to 1,100°F) yields an oil mixture that can be converted into phenols. At an estimated full-production cost of under $0.66/kg ($0.30/pound), they would be less expensive than phenols made from petroleum, and for this reason a government-industry consortium has formed to develop this into a commercial process. Although niche applications such as this have little impact on energy use, they could help establish a mature pyrolysis industry, which could then lead to eventual large-scale production of biofuels (Chum 1990). Pyrolysis oils could also be used to fire combustion turbines, an alternative to the integrated gasification/combustion turbine approach discussed earlier (DOE 1991a).

Liquefaction

Liquefaction, the process furthest from commercialization, usually involves taking the products of either gasification or pyrolysis and converting them, through catalytic reactions, to liquid fuels. Medium-Btu gas can be converted in this manner to methanol (methyl alcohol), a high-octane fuel commonly used in racing cars. With improvements in gasification technology, catalytic removal of tars and other noxious compounds from the hot gas, and the construction of large-scale plants producing up to 5,000 tons per day, it is estimated that methanol could be produced at a cost of about $0.60 per gallon (Chem Systems 1989). This would make it roughly competitive with gasoline at a wholesale price of $1.05 a gallon, or a retail price of $1.50 a gallon when mark-up, distribution costs, and taxes are considered.[8]

Methanol and its cousin, ethanol, can be burned in pure form in conventional automobile engines, although some modifications are required because of the fuels' corrosive effects on rubber and plastic parts. Cars designed specifically to run on pure methanol and ethanol could gain an immediate 10-20 percent benefit in thermal efficiency over cars designed for gasoline, because the higher octane rating of alcohol fuels (about 110 compared to 90 or so for gasoline) allows engines to have a higher compression ratio (the degree to which air and fuel in the cylinders is compressed before ignition). Methanol- and ethanol-powered cars could also have higher acceleration than comparable gasoline-powered cars (which is why methanol is favored in racing cars). Without modifications to the fuel or engines, however, the cars might also be more difficult to start in cold weather because of the lower volatility (ease of evaporation) of alcohol fuels. In addition, they would need larger fuel tanks because of the lower volumetric energy content of alcohol fuels (Gordon 1991).

Biochemical Conversion: Biogas and Ethanol

Ethanol

One of civilization's earliest discoveries was that watery "mashes" of corn, potatoes, barley, grapes, or other food crops could be fermented to make ethanol (ethyl alcohol), or as it is more commonly known, grain alcohol. Though the cause of this transformation remained a mystery until the 19th century, when Louis Pasteur proved that yeast, a type of microscopic fungus, was the active agent, brewing, distilling, and drinking spirits had long been popular pursuits. For a time early in this century, ethanol competed with gasoline as an automotive fuel, and Henry Ford ardently supported its use in cars, going so far as to equip some early Model Ts with engines that could run on ethanol or gasoline and to lobby for increased alcohol fuel production during World War I (Sperling 1988).

Fuel ethanol again emerged as an industry in the early eighties. With the benefit of federal and state tax breaks equivalent to about $0.60 per gallon, sales of ethanol produced from grain surpluses jumped from 20 million gallons in 1979 to 850 million gallons in 1990. Ethanol is not sold to motorists in pure form, but in a 1:9 ethanol-gasoline blend called gasohol. About 8 percent of gasoline sold in the United States contains ethanol, and in some farm states gasohol has penetrated as much as one-fourth to one-third of the gasoline market. The U.S. ethanol industry, though substantial, is dwarfed by that of Brazil, which produces more than four billion gallons each year from sugarcane, and where most new cars are capable of burning pure ethanol (Gordon 1991).

Figure 5.4
Transportation may be the most important future application of biofuels because they can be burned in today's cars and trucks with only minor vehicle modifications. This Ford has been converted to burning any mix of gasoline, ethanol, and methanol. Source: Ford Motor Company.

At present, corn is the most important feedstock for ethanol production in the United States because its high starch content makes fermentation relatively easy, and because in most years millions of tons of surplus corn are available at a low price. Grain surpluses are a limited and variable resource, however, and, as we will see, the corn-ethanol conversion process is highly inefficient on a net energy basis. Consequently, researchers are investigating ways to convert wood and other potential energy crops to ethanol using more efficient processes (Hinman 1990, DOE 1990a). The basic problem is that common yeast cannot ferment lignocellulosic material, which makes up more than 90 percent of woody plants. So in order to convert the bulk of biomass into alcohol, this material must first be broken down into simple sugars through hydrolysis — a process in which hydrogen in water combines with complex organic compounds to form simpler sugars, starches and other carbohydrates. The two types of hydrolysis, acid and enzymatic, can be quite efficient but have different drawbacks. Acid hydrolysis, for example, tends to degrade some of the sugars themselves, decreasing the overall conversion yield, while commercial enzymes for the enzymatic process are expensive and ineffective against some compounds.

The discovery of a type of yeast capable of fermenting the five-carbon sugars in hemicellulose, which forms up to 20 percent of plant material, along with methods for converting unfermentable sugars into ones that can be digested, has dramatically increased the potential alcohol yield from biomass. Using genetic techniques, researchers are also creating yeast that can continue working in relatively high alcohol concentrations, thus increasing the yield further. The next step will be to combine both saccharification — making simple sugars from cellulose and other long-chain molecules — and fermentation into a single process, known as simultaneous saccharification and fermentation (SSF).

Lignin, a fibrous sugar polymer that cements plant cells together and gives a plant its structural integrity, is impervious to all these techniques, but thermochemical methods can convert it into other refined fuels that can be used to provide energy for the ethanol-production process or sold separately as a gasoline substitute.

With more than 90 percent of plant components theoretically usable, ethanol could be produced today from lignocellulosic biomass (that is, wood, grasses, and even waste paper), at an estimated production cost of \$1.20 to \$1.35 per gallon. By improving the conversion of cellulose to usable sugars, as described above, and decreasing the cost of the feedstock through energy crop cultivation, the Department of Energy expects this cost to drop as low as $0.60 to $0.80 per gallon by the end of the nineties. At this cost, and considering its energy content relative to gasoline, ethanol would be competitive with gasoline at a wholesale price of $0.85 to $1.15 a gallon,[9] or a retail price of $1.30 to $1.60 a gallon. This is approximately the range of gasoline prices currently projected for the late nineties (EIA 1991a).

One often-raised question is whether ethanol requires as much or more energy to produce (in cultivating the feedstock, storing and transporting it to the conversion plant, and converting it to ethanol) as is contained in the fuel. If so, then ethanol (and perhaps other biofuels) would not be a sensible energy alternative for the United States. The corn-ethanol process is notoriously inefficient, with a net "energy return on investment" (EROI) — energy output divided by energy input — estimated as low as 0.6, if the value of coproducts (such as animal feeds) is ignored (Pimentel 1991). The energy balance appears likely to be much more favorable for high-productivity conversion methods using dedicated energy crops such as short-rotation trees or grasses, however. The EROI for these processes is estimated to be roughly 5, that is, the external energy inputs required for ethanol production will equal about 20 percent of the energy contained in the ethanol. Producing the feedstock alone, without converting it to fuels, will require energy inputs equal to 13 to 18 percent of the energy output (Lynd et al. 1991).

Biogas

Another familiar biochemical process with significant energy potential is anaerobic digestion. In the absence of oxygen, microbes break down organic matter and release a medium-Btu mixture of methane and carbon dioxide called biogas. In landfills, this gas is often vented and flared to reduce local pollution and prevent a buildup of the potentially explosive gas. In a growing number of cases, however, the gas is being used to produce power. As of 1990, 117 landfill methane power plants were in operation, each a few megawatts in size, with 12 more planned (Beranyi 1991). Most of these plants use the gas on-site to generate electricity, though it can also be purified and sold to natural-gas distributors. Animal manure, crop residues, sewage, and industrial wastes can also be converted to energy through anaerobic digestion. About 300 small digestors were installed in the United States in the seventies and eighties and several million have been installed in developing countries, though how many of these are still in operation is a mystery (Klass 1988, Finneran 1986).

In general, anaerobic digestion is economical for on-site power production wherever sufficient wastes are available and there is a ready demand for the power. The cost of purifying the gas to remove both carbon dioxide and residues from the digestion process often makes it too expensive to be sold to natural-gas distributors, however. To make pipeline-quality gas less expensive, government-sponsored research has focused on improving the yield and reducing the cost of digestion systems (called reactors). In 1989, this research led to a new reactor that could generate methane in the presence of solids at a concentration of 30 percent, as opposed to the 10 percent achieved previously, implying that the new reactor can generate as much methane as a conventional reactor three times its size. At Walt Disney World near Orlando, Florida, another experimental digestor (jointly operated by the federal government, the Gas Research Institute, and the state of Florida) produces a pipeline-quality gas that is up to 95 percent methane (DOE 1990a, Biologue 1991).

Scientists have only recently begun to understand in detail the biological processes involved in anaerobic digestion, and experiments are being conducted to develop genetically modified organisms with improved performance. The sheer number of organisms involved and the complexity of interactions between them make progress slow, however. The Department of Energy estimates that the cost of producing pure methane in digestion systems has been reduced from $8/GJ in 1980 to about $4.50/GJ today; the goal is to reach $3.50/GJ by the mid-nineties (DOE 1990a).

Environmental Issues

More than most renewable energy sources, biomass raises important environmental issues. Will the possible benefits associated with its production and use on a large scale outweigh concerns over land impacts and pollution? Not enough information is available to provide a definite answer to this question, and much will depend on how carefully the resource is managed. The picture is further complicated because there is no single biomass technology, but rather a wide variety of production and conversion methods, each with a distinct set of environmental impacts. We can only touch on some of the areas of concern here.

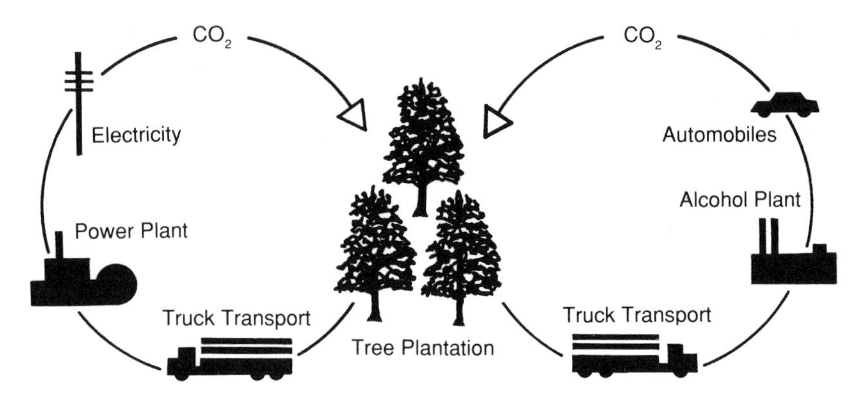

Figure 5.5
When biomass is produced and consumed for energy in a sustainable fashion, about the same amount of carbon dioxide generated in combustion is reabsorbed in plant growth, so there is no net contribution to greenhouse warming. This figure shows two possible fuel cycles, one supplying transportation fuels, the other electricity. Source: Weinberg and Williams (1990).

Air Pollution

Inevitably, the combustion of biomass or biofuels produces air pollutants, including carbon monoxide, nitrogen oxides, and particulates such as soot and ash. The amount of pollution emitted per unit of energy generated varies widely by technology, with wood-burning stoves and fireplaces generally the worst offenders. In cities such as Albuquerque, New Mexico, and Denver, Colorado, where many residents burn wood for home heating, winter restrictions on wood use have been established because of the visible effects on air quality. Modern, enclosed fireplaces and wood stoves pollute much less than traditional, open fireplaces for the simple reason that they are more efficient. Specialized pollution control devices such as electrostatic precipitators (to remove particulates) are available, but without specific regulation to enforce their use it is doubtful they will be widely adopted.

Emissions from conventional biomass-fueled power plants are generally similar to emissions from coal-fired power plants, with the notable difference that biomass facilities produce very little sulfur dioxide and toxic metals. The most serious problem is their particulate emissions, which must be controlled with devices such as cyclone precipitators.

More advanced technologies such as the whole-tree burner (which has three successive combustion stages) and the gasifier/combustion turbine combination should generate much lower emissions, perhaps comparable to those of power plants fueled by natural gas.

Facilities that burn raw municipal solid waste present a unique pollution-control problem as this waste often contains toxic metals, chlorinated compounds, plastics, and other materials that can generate harmful emissions. The problem is much less severe in facilities burning RDF, however, so that most plants built in the future will probably use this fuel. Co-firing RDF in coal-fired power plants may provide an inexpensive way to reduce coal emissions without having to build new power plants.

Using biomass-derived methanol and ethanol in vehicles could substantially reduce some types of air pollution. Both fuels evaporate more slowly than gasoline, thus helping to reduce evaporative emissions of volatile organic compounds (VOCs), which react with heat and sunlight to generate ozone. According to Environmental Protection Agency estimates, in cars specifically designed to burn pure methanol or ethanol, VOC emissions from the tailpipe could be reduced 85 to 95 percent, while carbon-monoxide emissions could be reduced 30 to 90 percent. In flexible-fueled cars that can burn gasoline and alcohol in any mixture, the benefits of ethanol or methanol use would not be as large. In either case, emissions of nitrogen oxides, a source of acid precipitation, would not change significantly compared to gasoline-powered vehicles (Lynd 1989).

Some studies have indicated that the use of fuel alcohol increases emissions of formaldehyde and other aldehydes, compounds that have been identified as potential carcinogens. Critics counter that these results consider only tailpipe emissions, overlooking another significant pathway of aldehyde formation, VOCs, which are much lower in alcohol-burning vehicles (Alson 1990). A model of ozone formation in the Los Angeles area, where ozone levels exceed federal standards almost half the days in the year, shows that the introduction of vehicles burning pure methanol would decrease peak ozone concentrations up to 22 percent, with a minimal impact on aldehyde levels (Russell, St. Pierre, and Milford 1990). Overall, however, alcohol-fueled cars offer no real solution for air pollution in dense urban areas, where electric vehicles are far more likely to be favored (see chapter 8).

Thermochemical conversion of biomass can produce air pollutants, such as carbon monoxide and VOCs. Standard pollution control technology, however, can neutralize or remove most of these before they are released into the atmosphere. Pyrolysis and liquefaction also generate liquid and solid hazardous wastes such as tars, catalysts, acids, char, and ash that must be properly disposed of. This problem has received scant attention, however, and its implications for the incipient biofuels industry are consequently difficult to assess.

Greenhouse Gases

One point that is clear is that the substitution of biomass for fossil fuels, if done in a sustainable fashion, would greatly reduce emissions of greenhouse gases. The amount of carbon dioxide released when biomass is burned is very nearly the same as the amount required to replenish the plants grown to produce the biomass. Thus, in a sustainable fuel cycle, there would be no net emissions of carbon dioxide. As noted earlier, some fossil-fuel inputs would be required for planting, harvesting, transportation, and processing. Yet, if efficient cultivation and conversion processes are used, the resulting emissions should be small (around 20 percent of the emissions created by fossil fuels alone, based on estimates of the energy return on investment cited earlier). And if the energy needed to produce and process biomass came from renewable sources in the first place, the net contribution to global warming would be zero.

Similarly, if biomass wastes such as crop residues or municipal solid wastes are used for energy, there should be few or no net greenhouse-gas emissions. There might even be a slight greenhouse benefit in some cases, since methane is formed through anaerobic decay of biomass in landfills, and it is a more potent greenhouse gas than carbon dioxide.

Implications for Agriculture and Forestry

One perhaps surprising side effect of growing trees and other plants for energy is that it could benefit both soil quality and farm economies. Energy crops could provide a steady supplemental income for farmers in off-seasons or allow them to work some unused land without requiring much additional equipment. Moreover, energy crops could be used to stabilize cropland or rangeland prone to erosion and flooding. Trees would be grown for several years before being harvested, and their roots and leaf litter could help stabilize the soil. The planting of coppicing, or self-regenerating, varieties would minimize the need for disruptive tilling and planting. Perennial grasses harvested like hay could play a similar role; soil losses with a crop such as switchgrass, for example, would be far less than for annual crops such as corn (Tyson 1990).

If improperly managed, however, energy farming could have harmful environmental impacts as well. While energy crops could be grown with less pesticide and fertilizer than conventional food crops, large-scale energy farming — at least as it is now envisioned — could nevertheless lead to substantial increases in chemical use simply because more land would be under cultivation. It could also affect biodiversity through the destruction of species habitats if forests are more intensively managed or conservation reserve land is converted to monoculture energy crops. If agricultural or forestry wastes and residues were used for fuel, then soils could be depleted of organic content and nutrients unless care was taken to leave enough wastes behind. These concerns point up the need for regulation and monitoring of energy crop development and waste use (Cook, Beyea, and Keeler 1991).

Energy farms may present a perfect opportunity to promote low-impact sustainable agriculture, or, as it is sometimes called, organic farming. A relatively new federal effort for food crops is under way that emphasizes crop rotation, integrated pest management, and sound soil husbandry to increase profits and improve long-term productivity. These methods could be adapted to energy farming. Nitrogen-fixing crops could be used to provide natural fertilizer, while crop diversity and use of pest parasites and predators could reduce pesticide use. Though such practices may not produce as high a yield per hectare as more intensive methods, this penalty could be offset by reduced energy and chemical costs (Reganold, Papendick, and Farr 1990).

Increasing the amount of forest wood harvested for energy could have both positive and negative effects. On one hand, it could provide an incentive for the forest-products industry to manage its resources more efficiently, and thus improve forest health. But it could also provide an excuse—under the "green" mantle—to exploit forests in an unsustainable fashion. Unfortunately, commercial forests have not always been soundly managed, and many people view with alarm the prospect of increased wood cutting. Their concerns can only be met by tighter state and federal controls on forestry practices, following the principles of "excellent" forestry advocated by Robinson (1988) and others. If such principles are applied, it should be possible to extract energy from forests indefinitely.

Notes

1. In 1987, industrial roundwood production in the United States (excluding logging residues) was 14.5 billion cubic feet (Bureau of Census 1990). Assuming an average oven-dried wood density of 30 pounds per cubic foot (USDA 1991d), this is equivalent to about 200 million metric dry tons of biomass. The heating value of oven-dried hardwood averages about 20 GJ per metric ton (EIA 1991b).

2. This number assumes an average production of 70 cubic feet per acre (5 cubic meters per hectare) and average wood density of 30 pounds per cubic foot (0.48 metric tons per cubic meter).

3. This assumes an average biomass yield of 30 dry metric tons per hectare per year, energy content of 20 GJ per ton, and an efficiency of conversion to biofuel of 50 percent.

4. These figures are drawn from Tyson's "unrestricted land" case. The study also considered a "restricted land" case, in which it was assumed that biomass crops would be confined to marginal cropland. In this case, biomass production costs would increase 10 percent and land requirements 40 percent.

5. The plant size is dictated largely by the distance over which wood chips can be economically collected and hauled by truck, which is around 75-100 miles (CEC 1987).

6. Three types of gas turbine (similar to jet engines) could be used with successively higher efficiencies and cost: the ordinary gas turbine, the steam-injected gas turbine, and the intercooled steam-injected gas turbine. A competing concept is the combined cycle system, which consists of a gas turbine and steam turbine operating jointly to maximize efficiency.

7. In fiscal 1991, the federal government appropriated $391 million to clean coal technologies but only $33 million to biofuels technologies. See Appendix B.

8. The heating values of methanol and gasoline are 66,000 Btu per gallon (19 MJ/l) and 124,000 Btu per gallon (35 MJ/l), respectively (DOE 1988b). This implies that 1 gallon of gasoline is equivalent on an energy basis to 1.8 gallons of methanol.

9. Ethanol's heating value is 86,600 Btu per gallon (24 GJ/l), implying that a gallon of gasoline is equivalent to 1.4 gallons of ethanol. Because of efficiency improvements possible in cars designed optimally for alcohol fuels, however, the ratio could be as low as 1:1.25 (Wyman and Hinman 1990).

6 Energy from Rivers and Oceans

Photographs of the earth taken from space leave little doubt that ours is a water planet. Massive oceans separate the smaller continents, each crossed by a network of rivers and dotted with lakes. For millennia, people have used this water as a source of food, transportation, and energy. Water collects energy in several ways — it stores solar energy in the warm surface layers of oceans and lakes; it evaporates into the atmosphere and falls as rain or snow, creating rivers; it stores kinetic energy in ocean currents; and it even captures gravitational energy from the moon and sun as tides.

Not surprisingly, a variety of methods has been devised to capture these stored energies and convert them into a useful form. One of the four technologies discussed here, hydropower, is already in wide use, supplying 10 to 12 percent of U.S. and around 20 percent of world electricity demand. The other three, ocean thermal energy conversion (OTEC), tidal power, and wave power, have been explored to varying degrees, although none has yet been developed on a significant scale. Of the latter, only OTEC offers at least the possibility of having a significant impact on energy supply in the United States, although the others could find important applications elsewhere.[1]

Hydropower

Although it was coal that pushed the Industrial Revolution into its maturity in England, water power attended its birth. In the textile industry, for example, 18th-century inventors used water wheels to run spinning jennies, shuttles, and looms. Because fast-running rivers were often located far from major markets or trading ports and their flows varied with the season and amount of rainfall, however, they were often an inconvenient power source. Adapting textiles and other industries to coal- and wood-fired engines allowed industry to move to cities, closer to both markets and imported raw materials, leaving water power largely behind.

Rivers played a crucial role in the expanded European settlement of the United States. The Ohio, Mississippi, and Missouri Rivers, among others, carried settlers into the heart of the country and brought natural resources found there back to coastal manufacturing and trading centers. Thus it is natural that hydroelectric power, or hydropower, became one of the most important sources of electricity in the early part of this century, after the first hydropower plant was completed at Niagara Falls in 1878. Even today, rivers provide 10 to 12 percent of U.S. electricity supply (270 to 320 billion kilowatt-hours), depending on demand and rainfall, or as much as is produced by 40 to 50 nuclear power plants. In

Figure 6.1
The 1,455 MW Hoover Dam is typical of large hydroelectric plants built in the United States in the early and middle part of the century. Source: U.S. Department of Energy.

a good year, hydropower displaces around 3.4 EJ of primary fossil energy, or about 4 percent of total primary energy demand (EIA 1991b).

The Resource

In principle, there remains much room for further hydropower development in the United States. The Federal Energy Regulatory Commission (FERC) has cataloged 7,243 sites, which, based on estimates of natural stream flows, could support 147,000 MW of hydropower capacity. As of 1991, only 2,279 sites with 73,000 MW total capacity had been developed, with another 86 projects of 3,300 MW capacity planned or under construction (most of which were expansions or upgrades of existing facilities). Thus, the potential exists for the United States to just about double its current hydropower capacity. The majority of this expansion potential resides in western states, where most previous hydropower development has taken place (FERC 1988, 1991).

Economic, regulatory, and other constraints are likely to block the development of most of this resource, however. Environmental laws, in particular, place severe restrictions on new hydropower development. The 1968 National Wild and Scenic Rivers Act and other federal legislation preclude building facilities on stretches of many virgin rivers, eliminating about 40 percent of the nation's remaining undeveloped resource. An additional 19 percent of potential sites are under a development moratorium until their final status can be decided. All told it seems likely that less than half the remaining potential — perhaps 30,000 MW — will actually be available for development. What is more, according to a 1990 report by national laboratory scientists, only 22,000 MW of the undeveloped hydropower resource is economically viable, and of this only 8,000 MW is likely actually to be developed because of "regulatory complexities and institutional and jurisdictional overlaps" in the hydropower licensing process (NREL 1990b). These regulatory issues are discussed later.

Many of the opportunities for expanding hydropower consequently involve upgrading or expanding existing facilities rather than building new ones. The capacity of the immense (6,180 MW) Grand Coulee Dam, for example, was boosted in the eighties by more than 2,000 MW with the addition of three new turbines and the rewinding of existing turbines. Several federal hydropower projects were also uprated in the past decade, resulting in a 1,137 MW increase in total capacity (Bureau of Reclamation 1991). It should be kept in mind, however, that expanding the peak capacity of a hydropower plant does not necessarily increase its annual generation, as that is limited by total annual streamflow.

Proponents of small-scale hydropower point out that there are 70,000 or so nonhydroelectric dams intended for irrigation, navigation, and flood control that could be rigged for electricity generation (Flynn 1991). The large number of nonhydro dams may be misleading, however, as it is hard to say what fraction of these could be converted or refurbished at a reasonable cost. If one study is any indication, the fraction is probably very small: Out of an inventory of over 10,000 dams in New England, only 320 were identified that could be retrofitted to generate electricity at a levelized cost of 19¢/kWh or less. The potential power output of these sites was just 300 to 600 MW (New England River Basins Commission 1981).

A more promising option may be to refurbish some of the more than 3,000 small hydropower facilities abandoned in the fifties and sixties. Cornell University, for example, recently refurbished a small hydropower plant that was closed in 1969. Renovations to the 1.3 MW plant, including new turbines, cost $1.25 million. It now supplies 5 percent of Cornell's electricity needs and sells $250,000 worth of electricity each year to the New York State Power Authority (Dowling 1991).

Pumped hydropower storage is often included in statistics on national hydropower capacity and potential, although it actually only stores electricity produced by other facilities by pumping water from a lower reservoir to a higher one. Some hydropower plants both store electricity and produce it on their own; others are dedicated pumped-storage plants. As of 1988, 17,000 MW of pumped-storage capacity were in operation (FERC 1988). (For a complete discussion of this technology, see chapter 8.)

Technology and Costs

Hydropower plants, both large and small, are a proven technology. In large plants, a reservoir is created by damming a river, and the water in the reservoir is then allowed to fall up to hundreds of feet through a turbine, generating electricity. (As a rough rule of thumb, one gallon of water per second falling one hundred feet can generate one kilowatt of electrical power.)

Designs for small hydropower plants (typically less than 30 MW capacity) vary considerably. Some have dams with a head (the distance between the top of the reservoir and the bottom of the dam) of less than

60 feet. Others do not have dams at all, but instead channel some or all of the streamflow directly through turbines located in midstream or offstream. By avoiding the creation of a reservoir, such axial-flow designs can have a smaller relative impact on river ecosystems both upstream and downstream.

Traditionally, hydropower has been one of the least expensive sources of electricity. Some of the immense facilities that were built in the heyday of hydropower construction from the thirties to the sixties generate power today at a cost of under 1¢/kWh. The cost of new hydropower plants varies widely, depending on such factors as the size of the plant, the design, the capacity factor (typically 40 to 50 percent, as hydropower plants are most often used to provide intermediate and peak, not baseload, electricity), licensing delays, environmental mitigation costs, and proximity to transmission lines. As is the case with most renewable energy sources, initial capital costs are relatively high, whereas operations and maintenance costs are low. Large plants typically cost between $500 and $2,500 per installed kilowatt of capacity, while the cost of small plants ranges from less than $1,000 to more than $6,000 per kilowatt and averages around $2,000 per kilowatt. Uprating the capacity of an existing plant costs much less than building a new plant, usually less than $100 per kilowatt of increased capacity. Operations and maintenance costs average 0.2¢/kWh at privately owned plants, and about half as much at government-owned plants (Shea 1988, Bureau of Reclamation 1991, EIA 1990).

From a utility's perspective, a key advantage of hydropower is that it can easily be switched on and off, making it ideal for meeting peak and emergency demand. It is also inexpensive to operate, so it can be used as a baseload source where sufficient streamflow exists. On the other hand, hydropower generation is strongly affected by the seasons and precipitation. The drought of 1988 caused a 25 percent drop in national hydropower output, and relatively snowless winters in the West have reduced power output from some hydropower plants there in recent years. And if predictions of increased summer dryness caused by global warming prove correct, water resources may become more scarce in the future, leading to further reductions in hydropower generation.

Environmental and Regulatory Issues

Despite the apparent economic and practical advantages of hydropower, its development has become increasingly problematic. The construction of large plants has virtually ceased because most suitable undeveloped sites are under federal protection. To some extent, the slack has been taken up by a revival of small-scale development sparked by the Public Utilities Regulatory Policy Act (PURPA, discussed in chapter 3) and federal tax credits. But small-scale hydro development has not met early expectations. Between 1984 and 1988, only 650 MW of small-scale hydro capacity were added. As of 1988, small hydropower plants accounted for 7,235 MW, or about 10 percent, of total hydropower capacity (FERC 1988).

Declining fossil-fuel prices and reductions in renewable energy tax credits are only partly responsible for the slowdown in hydropower development. Just as significant have been the effects of public opposition to new development and the many environmental regulations that affect it. According to the industry view, current regulations place an excessive burden on developers, adding 30 percent to the average cost of new small facilities. Moreover, the 1986 Electric Consumers Protection Act (ECPA), which was designed to place consideration of fish and wildlife interests on an equal footing with power generation, has raised such stringent environmental requirements that it has, in the words of one industry representative, "ensured that new small hydro will not be built" (Rogers 1989).

Perceived problems such as these prompted the Bush administration to propose, in its 1991 National Energy Strategy, that hydropower licensing procedures be consolidated and streamlined (NES 1991). As it has taken shape in legislation, this proposal would allow states to take over regulation of hydropower projects of 5 MW size or less (about two-thirds of licensed projects). It would also restrict the role of federal agencies other than FERC (such as the Forest Service) in setting conditions for hydropower projects affecting public lands. Needless to say, this proposal has been loudly opposed by environmental groups, a representative of one of which testified to Congress that the change in rules "could lead to the dewatering of thousands of miles of rivers and streams and major damage to fisheries and wildlife" (Conrad 1991).

Environmental regulations affect not only new projects but existing ones, as well. For example, a series of large facilities on the Columbia River in Washington will probably be forced to reduce their peak output by 1,000 MW to save an endangered species of salmon, whose numbers have declined rapidly because the young are forced to make a long and arduous trip downstream through several power plants, risking death from turbine blades at each stage. To ease this trip, the hydropower plants may be required to "flush through" — divert water around their turbines — at those times of the year when the fish attempt the trip. And in New England and the Northwest, there is a growing popular movement to dismantle small hydropower plants in an attempt to restore native trout and salmon populations.

That environmental concerns would constrain hydropower development in the United States is perhaps ironic, since these plants produce no air pollution or greenhouse gases. Yet as the salmon example makes clear, they can have severe impacts on the environment. The effects of very large dams are so great that there is almost no chance that any more will be built in the United States, although large projects continue to be pursued in Canada (the largest at James Bay in Quebec) and in many developing countries. Depending on the location, the reservoirs created by such projects can inundate large areas of forest, farmland, wildlife habitats, and even towns. In addition, the dams can cause radical changes in river ecosystems both upstream and downstream. Sediments bearing nutrients accumulate in the reservoir instead of being carried downstream. The water flowing out of the reservoir generally has a higher temperature and lower dissolved oxygen content, which can alter the balance of plant and animal life. Reservoir evaporation increases salt and mineral content, a serious problem for the Colorado and other mineral-rich rivers. Power generation and pumped-storage operation can cause large fluctuations in downstream water flow. Finally, dams can block fish migration and destroy spawning grounds, resulting in losses for commercial fisheries and sport fishing.

Small hydropower plants that have reservoirs can cause similar types of damage, though obviously on a smaller scale. Some of the impacts of both large and small facilities can be mitigated by installing "ladders" or other devices to allow fish to migrate over dams, and by maintaining minimum river-flow rates; screens can also be installed to keep fish away from turbine blades. In one case, flashing underwater lights placed in the Susquehanna River are helping direct night-migrating American shad around turbines at the York Haven Hydroelectric Station in Pennsylvania (Caruso 1991). As environmental regulations have become more stringent, developing cost effective mitigation measures such as this one has become a priority for the electric power industry.

Despite these efforts, however, hydropower is almost certainly approaching the limit of its potential in the United States. In one of the most optimistic projections, the Bush administration's National Energy Strategy predicts that up to 16,000 MW of new and refurbished capacity could be added to the 73,000 MW now in operation over the next 40 years *if* proposed changes in environmental regulation take effect.[2] In reality, such a major overhaul of licensing procedures is unlikely, given the public's increasingly acute awareness of environmental issues. It is just as likely that total capacity and annual generation will decline over the long term because of increased demand on water resources for agriculture and drinking water, declining rainfall (perhaps caused by global warming), and efforts to protect or restore endangered fish and wildlife.

Over half of America's hydropower resource has already been developed. Inevitably, society will place a high value on preserving as much of the remaining resource as possible.

Ocean Thermal Energy Conversion

Oceans are the largest natural collectors of solar radiation. Covering more than 70 percent of the earth's surface, their upper layers are warmed to temperatures as high as 26°C in the tropics while their depths remain near freezing. This temperature differential can be used to run a heat engine to generate electricity, an idea first proposed by Jacques d'Arsonval of France in 1881. The first practical application of this idea was by another Frenchman, Georges Claude, who built a small ocean thermal energy conversion (OTEC) plant off the coast of Cuba in 1929. Despite continued investigation of OTEC by the United States, however, no commercial facilities have yet been built.

Needless to say, considering the oceans' vastness, the total amount of energy stored in thermal layers is immense. According to one estimate, as much as 10 million megawatts of electric power could be generated worldwide from OTEC (Wick and Schmitt 1981). OTEC systems cannot be sited just anywhere, however. The best sites are in those areas where the surface temperature is high and where depths are sufficient to reach near-freezing water. A minimum annual average temperature differential of 18°C is regarded as necessary for OTEC operation,[3] a requirement satisfied throughout much of the subtropical and tropical oceans (Penney and Bharathan 1987, Cohen 1982).

Three approaches have been suggested for extracting and marketing ocean-thermal energy. In the first, an OTEC power plant would be based onshore, and the cold and warm seawater would be piped in. In the second, the plant would be moored offshore, and the electricity generated would be transmitted through underwater cables. Although there are many densely populated coastal areas in the subtropical southern

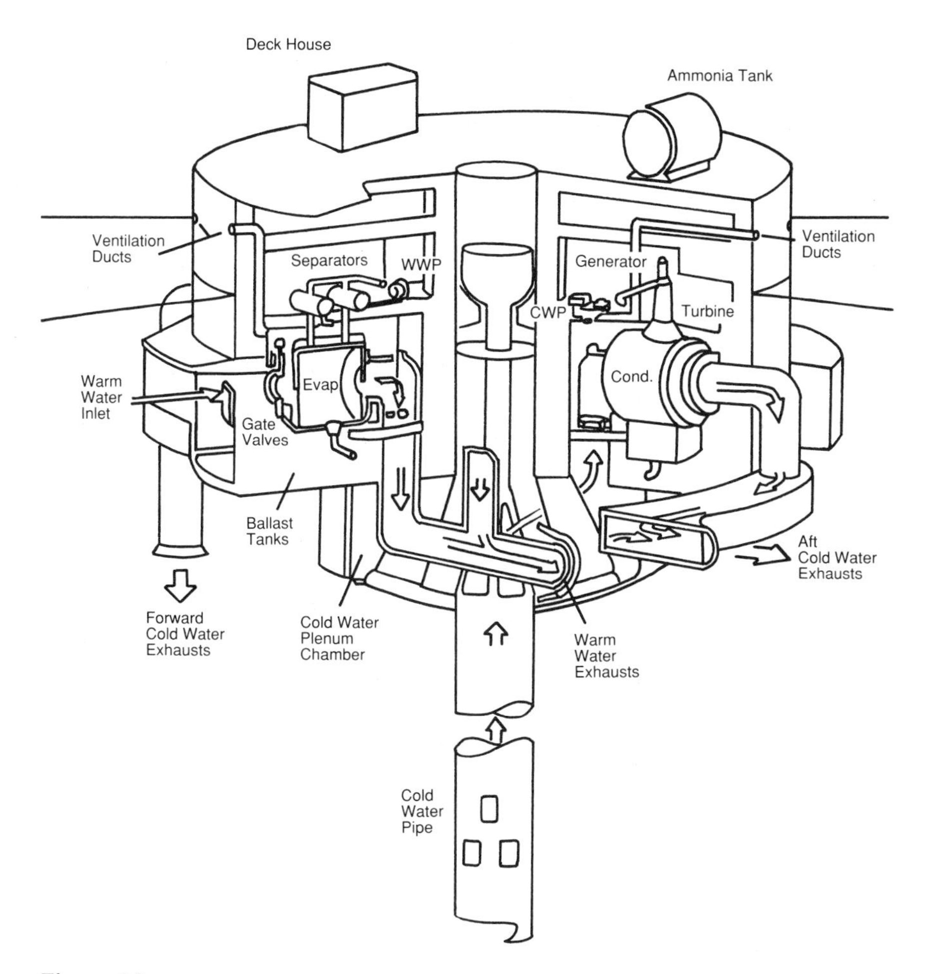

Figure 6.2
Schematic drawing of a 100 MW, closed-cycle OTEC plant designed by TRW, Inc. Source: Cohen (1982).

United States and in other continents, ocean waters over the continental shelf are generally too shallow to permit OTEC operation over reasonable transmission or piping distances. Consequently, both onshore and moored OTEC systems will probably find their widest (and certainly earliest) application providing power to tropical island chains such as Hawaii, Indonesia, and the Philippines, which are mostly dependent on high-priced oil for their electricity. (Not coincidentally, Hawaii has been a leading funder of OTEC development since 1979.) Both of these types of OTEC system could produce valuable byproducts in addition to electricity, including fresh water, cold water for district cooling of buildings, and nutrients and minerals for marine culture.

In the third approach, the OTEC system would operate on an unmoored ship, and the electricity generated would be used on-site to manufacture hydrogen (or possibly high-value, energy-intensive products such as ammonia), which could then be shipped to markets. This scheme would free OTEC plants to roam at will, "grazing" the oceans for the best thermal gradients and thereby opening up vast potential for energy production (Cohen 1982). Like other ideas for generating hydrogen from renewable electricity sources, however, this will not become practical until the cost of OTEC-generated electricity is reduced to a very low level, perhaps 2¢/kWh (see chapter 8).

Technology and Cost

Engineers seeking to design a cost effective OTEC system immediately face three major challenges. First, because of the relatively small temperature differentials between ocean surface and deep layers, the ideal (Carnot) thermal efficiency of the OTEC power cycle is inherently very low, around 6 or 7 percent. This implies that some critical components such as heat exchangers (required for transferring heat from seawater to a working fluid and back again) must be large and probably expensive. Second, large volumes of seawater must be circulated through the system, and the electricity required for this reduces the net plant efficiency still further. And third, OTEC components that come into contact with seawater must be protected from corrosion and biofouling (the accumulation of sealife). Much of the history of OTEC development has involved investigating ways to overcome these challenges.

The power-generating component of OTEC systems comes in two basic types, *closed cycle* and *open cycle*. In closed-cycle plants, the heat from warm surface water is transferred to a working fluid such as ammonia or a fluorocarbon, which is vaporized and then used to turn an electricity-generating turbine. Cold water pumped from the ocean depths recondenses the vapor, completing the cycle. The main disadvantage of this design is that it requires enormous heat exchangers for fluid evaporation and condensation. These could make up 30 percent of the cost of the entire plant (DOE 1990d).

Open-cycle plants, such as the one used by Claude in his 1929 experiment, use seawater itself as the working fluid, flashing it to steam in a depressurized chamber. Although open-cycle power plants have no heat exchanger,[4] they require steam turbines capable of operating at very low pressures, which could be even more expensive.

In most other respects, the two designs are very similar. Both require pipes ranging from a few meters to several tens of meters in diameter, and about a kilometer long, made from aluminum, polyethylene, or even reinforced concrete for drawing cold water up from the ocean depths. Because the net thermal efficiency of OTEC is so low — around 2.5 percent, including parasitic pumping requirements and thermal losses in heat exchangers (Cohen 1982) — huge volumes of water (around a thousand gallons per second for each megawatt of capacity) are needed to drive the system. The cold water pipes must be able to withstand stresses from ocean currents, and methods must be developed for deploying them in the open ocean.

Offshore OTEC systems require stable platforms (floating or tethered) able to survive the rigors of storms. Much of this technology has already been developed for the offshore oil industry, but feasible depths for mooring are limited to about 2 kilometers. In addition, if hydrogen or other products are not manufactured on-site, offshore plants will require ship-to-shore power cables laid at depths two or three times greater than depths at which such cables have been laid in the past, and capable of withstanding constant flexing from the moving platform (Sanders 1991, Cohen 1982).

It should be clear by now that OTEC technology presents several unique engineering challenges not encountered in any other energy technology. Whether these challenges can be overcome in a commercial-size system remains to be seen. Unfortunately, actual operating experience with OTEC systems has been extremely limited. In the United States, two modern, small-scale OTEC demonstration plants were tested off the coast of Hawaii from 1979 to 1981, one a 50 kW plant called Mini-OTEC, the other a 1 MW plant called OTEC-1. Only one other plant, a 120 kW system installed by Japan on the Pacific island of Nauru in 1981, has ever been tested.

Mini-OTEC and OTEC-1, both closed-cycle systems, provided some answers to questions that had been raised, but also left important issues still to be resolved. Among other things, mini-OTEC demonstrated the feasibility of producing net power from an OTEC system — an important step, since Claude's experiment had consumed more electricity for pumping and other needs than the 22 kilowatts produced by the turbine. Mini-OTEC also demonstrated the ability of chlorination to prevent biofouling for at least a limited time on heat-exchange surfaces. (Subsequent experimentation in Hawaii has demonstrated that intermittent chlorination prevents biofouling.) OTEC-1 was sized for 1 MW capacity but was not equipped with a turbine and hence did not generate electricity. Rather, it was designed to test heat exchangers, pumps, and other components under ocean conditions. Unfortunately, because of cuts in the federal ocean energy research budget, this experiment was stopped just four months after operations had begun, so only limited data on biofouling, corrosion, and other potential problems could be obtained (Cohen 1982).

Since then, there has been little in the way of demonstrable progress in the development of OTEC technology. (It is telling, for example, that most of the literature on OTEC dates from the seventies and early eighties.) The Department of Energy ocean energy research program was drastically cut in the early eighties and is perennially in danger of being shut down. It is only because of continued interest from the state of Hawaii that it survives at all, and then only barely with annual funding of about $2 million. This is in stark contrast to the grand plans enacted into law in 1980, President Carter's last year in office, which called for 100 MW of OTEC demonstration plants by 1986 and 500 MW by 1989, and which set an industry goal of 10,000 MW installed in the United States by 1999.

Because of the lack of practical experience, the only data on OTEC costs and performance come from engineering studies, some over a decade old. If these studies are realistic, then OTEC could supply electricity at a cost at least competitive with that seen on island chains such as Hawaii. The near-term goal of the federal ocean energy research program is to develop the technical knowledge to construct an OTEC system at a cost below $7,800 per kilowatt (1987 dollars). This cost is roughly equal to the current capital cost of photovoltaic systems, but because OTEC plants could produce a constant output 24 hours a day, the levelized cost of electricity would be much lower, perhaps 10-12¢/kWh. The Department of Energy's long-range goal is $3,500 per kilowatt for Gulf and Atlantic coastal areas (DOE 1990d). After experimenting with various components such as cold water pipes, heat exchangers, and low-pressure turbines, the Department of Energy has chosen to focus its research on developing the technology for relatively small (2 to 15 MW) open-cycle plants, at least in part because they could generate fresh water as a byproduct for sale on tropical islands.

Far more ambitious goals have been laid down by Sea Solar Power, a company started by one of the technology's modern pioneers, J. Hilbert Anderson. On the strength of engineering studies, this company claims to be able to build 100 MW OTEC power plants at an initial cost of $2,500 per kilowatt and to sell power at a levelized price of 6.5¢/kWh. Once the technology is mature, the company predicts an installed cost of $1,500 per kilowatt, less than half the Department of Energy's long-term goal (Nicholson 1990). The main innovations in the closed-cycle Sea Solar Power design appear to lie in the submerged heat exchangers, which are predicted to be about three times more efficient (and hence three times smaller in heat-transfer area) than those of previous conceptual designs (Anderson n.d.). Other experts have expressed doubt about Sea Solar's ability to meet these targets, however (Gormley 1988).

Two other companies, GEC Marconi Research Center and Alcan Aluminium, are jointly designing their own closed-cycle plant, which would use heat exchangers made of aluminum instead of the traditional, but much more expensive, titanium. In their heat-exchanger design, parallel aluminum plates are bonded together and seawater and the working fluid, ammonia, are passed through them (Fitzpatrick, Hron, and Hryb 1991).[5] The two companies predict that a complete OTEC power plant of 500 to 5,000 kilowatts capacity, intended for use in conjunction with a marine culture operation (so that the cold water pipe and pumps would not be included in the price of the power plant), would cost from $1,500 to $3,200 per kilowatt (Johnson 1991).

Environmental Issues

The main environmental issues associated with OTEC concern its effect on ocean temperature, chemistry, and life. If developed on a truly massive scale — a million megawatts or more worldwide — there could be a noticeable cooling of surface waters that could influence ocean currents and climate (Kendall and Nadis 1980). But this level of OTEC deployment is unlikely to be reached anytime soon, if ever. Of more serious concern are the local impacts of OTEC use, particularly around tropical islands. It is possible that, on a local scale, the temperature change and its secondary effects could be quite significant in very large installations (100 MW or more).[6] Moreover, by bringing up cold water from deep beneath the surface, OTEC plants would add large quantities of minerals and nutrients to surface waters, thus stimulating algae and fish growth. OTEC proponents cite these changes as an important benefit for marine culture, but there may be costs associated with it that have not been identified, such as changes in the natural local balance of flora and fauna. In sum, while there are no obvious impediments to developing and deploying this technology in the near term, further study is needed to determine its large-scale impacts.

Tidal Power

Both the sun and moon exert gravitational force on the earth's oceans, causing a slight bulge to develop that we see as tides. This bulge constantly changes with the moon's orbit and the earth's rotation, and along with sea-coast and sea-bottom topography, this accounts for the differences in tidal times and heights observed at various points along coastlines. In some places tidal changes are quite small, but in others they are amplified by local topography so that they reach a quite stunning magnitude. In Canada's Bay of Fundy, for example, the difference between high and low tide can be as much as 16 meters (50 feet) (Charlier 1982, Sanders 1991).[7]

Compared to other renewable water resources, however, the potential of tidal power is quite limited. Because of constraints of local geography and tidal patterns, there are only a few sites around the world where this technology is likely to prove economic. One of the sites with the largest potential is the Bay of Fundy, where it is estimated that 6,000 MW of tidal capacity could be installed. Another is the Severn estuary on the western side of England, which could support up to 12,000 MW. In the United States, there are only two promising sites, the Passamaquoddy Bay in Maine (off the Bay of Fundy) and the Cook Inlet near Anchorage, Alaska, with a combined potential of 3,600 MW. All told, about 3 million megawatts are dissipated in tides worldwide, of which it is estimated that no more than two percent (60,000 MW) could realistically be captured in tidal power plants (Charlier 1982).

Generating electricity from tides is very similar in concept to generating it from rivers. The rising tide enters a reservoir through sluice gates in a dam across the mouth of a tidal bay. At high tide, these gates shut, and the water is impounded until just before low tide, when the reservoir head is at a maximum. Water released from the dam is then passed through turbines to generate electricity. In comparison with conventional hydropower, the heads developed in tidal dams are relatively small and therefore require special, low-flow-rate turbines. More important, the times of high and low tide change from day to day and so do not regularly coincide with times of peak power demand. Power-leveling strategies, like pumping water into the basin after high tide, can help reduce these fluctuations and produce more constant power.

The principles of tidal power were proven practical three decades ago, when the oldest (and still largest) tidal power plant, a 240 MW facility at La Rance in northwest France, was put into operation in 1966. There is just one such plant in North America, a 20 MW pilot plant at Annapolis Royal, Nova Scotia, in the Bay of Fundy. The costs associated with developing tidal power depend strongly on the site. So little data are available, however, that it is difficult to identify even a plausible range.

The lack of experience with tidal power also means that its environmental impacts are not wholly understood. Tidal dams will tend to change the average amplitude of high and low tides in the surrounding region, but by how much is difficult to say with any precision since the detailed topography of the basin will strongly influence the tidal response. Modeling results for the Bay of Fundy have indicated that the response to a sizable facility would be modest. Nevertheless, since estuaries and coastal marshes tend to be prime spawning grounds for fish and shellfish, even small changes in the daily tides may have a significant impact on marine life. The rapid release of water from tidal dams during the power generation cycle may also increase nutrient mixing, but the effect this will have on plant and animal life is unknown.

In sum, tidal power is potentially a significant resource in only a few parts of the world. Where feasible, it is an option worth exploring, but important questions concerning its cost and impacts on tides and marine life remain to be answered.

Wave Power

The third method of extracting energy from the oceans involves capturing the mechanical energy of ocean waves, which are created by winds. Hundreds of wave-power schemes have been patented around the world, and a number of different approaches are considered possibilities for near-term commercial generation. Some use the up-and-down motion of waves, others their rocking motion, surging motion, and even the variations they cause in underwater pressure (McCormick 1981).

Wave-energy technology has not been widely developed for commercial application, however. On a very small scale, more than a thousand 60-watt Masuda buoys, which use an oscillating column of water to drive an air turbine, power navigational aids around the world. Two demonstration plants, a 500 kilowatt oscillating water column system and a 350 kilowatt tapered channel system (similar to a tidal power plant), are operating in Norway, and there have been plans to build larger facilities on the islands of Tonga in the South Pacific and Bali in Indonesia. Most other wave-power designs have only been tested on a prototype or model scale, including a 20 kilowatt system on Lake Michigan.

Cost of energy comparisons are difficult to make with such limited data, but according to one analysis, utility-connected systems ranging in size from 500 kilowatts to 750 megawatts could be built at a cost of $580 to $4,830 per kilowatt (1987 dollars), depending on the plant type, size, and location. The projected levelized cost of electricity lies in the range of 10 to 20¢/kWh (Hagerman and Heller 1988a, 1988b).

As far as the United States is concerned, wave energy is not likely to make a significant contribution. The most likely areas for development are in the North Atlantic, the Sea of Japan, and off the coast of Australia.

Notes

1. Two types of ocean energy are not discussed here: energy from ocean currents, and energy from salinity gradients. For information on these technologies, the reader should refer to U.S. Senate, Committee on Science and Technology, *Energy from the Ocean*, April 1978. Also see McCormick (1981).

2. This projection is the basis of the estimated total hydropower resource (4 EJ) cited in chapter 2.

3. If the temperature differential is lower, more electricity is required to operate the plant than is produced by it, thus making sustained operation impossible.

4. If the plant is to produce fresh water as a salable byproduct, however, a heat exchanger will still be needed for the condensing process to keep the seawater and fresh water separated.

5. This heat exchanger technology could also be used to improve the efficiency of conventional steam power plants by collecting some of the heat remaining in the warm water effluents from these plants (Cohen 1992).

6. Analysis suggests that a 20 MW OTEC plant would cause an ocean-surface temperature drop of 0.02°C to 0.07°C in its immediate vicinity (Dugger, Francis, and Avery 1978). By extrapolation, a 100 MW plant would cause a drop of 0.1°C to 0.35°C, a potentially significant change comparable to natural long-term temperature variations.

7. Just as a tile bathroom amplifies a person's voice when certain notes are sung, the Gulf of Maine, which feeds the Bay of Fundy, acts as a natural amplifier because its dimensions are approximately in resonance with the rhythm of the tides.

7 Geothermal Energy

Volcanic eruptions through the ages have presented visible proof of the immense amount of heat trapped in the earth's interior. Capable of more explosive energy than a nuclear bomb, these eruptions have been responsible for some of history's greatest natural disasters. Some historians believe that a tidal wave set off by an immense eruption sparked the beginning of the decline of the ancient Minoan civilization of Crete, in about 1500 B.C. In more recent times, lava and ash from the eruption of Mount Pelee engulfed Saint-Pierre, the capital city of Martinique, in 1902, leaving only one survivor out of 28,000 inhabitants.

Although the violent heat of volcanoes is of little practical use to humanity, gentler forms of geothermal energy have been used for centuries (Armstead 1983, Anderson and Lund 1987). The ancient Greeks, Etruscans, and Romans soaked in public baths that used hot water from natural springs. In the 18th and 19th centuries, the purported curative and prophylactic benefits of mineral-rich thermal springs made health spas popular in Europe and the United States. Even today, in countries such as Japan, Hungary, and Iceland, public facilities for bathing remain in common use. Hot springs have also been prized for the minerals and metals they bring to the surface, which can be extracted and used in industry.

Only in the last hundred years or so, however, have geothermal sources been exploited on a significant scale for energy. Electricity was produced from natural steam for the first time in Larderello, Italy, in 1904. No other country followed Italy's example until 1958, when New Zealand built a power plant in the Wairakei area of the North Island. The first such plant in the United States went into service in California in 1960, and today nearly 3,000 MW of geothermal-electric capacity are in operation in this country, most of it at The Geysers field in northern California.

Besides generating electricity, the earth's heat is also being used directly for heating and cooling buildings and in agriculture, aquaculture, and industry. Iceland has pioneered large-scale geothermal district heating, and today heats more than 80 percent of its houses and buildings from local geothermal sources. In the United States, geothermal space and district heating projects appeared around the turn of the century in Klamath Falls, Oregon, and Boise, Idaho. Applications to farming, aquaculture, and industry have since been developed as well.

Counting both electricity and direct use, geothermal energy currently supplies about 0.2 percent of the primary energy consumed in the United States, making it the third-ranking nonfossil and nonnuclear energy source after hydroelectric and biomass (EIA 1991a). There is little question that it could supply a much larger fraction of U.S. and world energy demand in the future. Estimates place the U.S. geothermal resource base — the energy theoretically accessible with modern drilling technology — in the millions of exajoules, or thousands of times larger than domestic reserves of coal (NAS 1987). Hot water is known to exist underground in hundreds of locations in western states but is not yet being exploited. In addition, other types of geothermal sources, particularly so-called hot dry rock, may be developed in the future using techniques that have already been demonstrated on a limited scale. If these new techniques fulfill their promise, enormous amounts of energy could be supplied from geothermal sources in most parts of the country, not just in the West.

The Geothermal Resource

Heat stored within the earth's crust originates mostly from radioactive decay of unstable elements such as uranium, potassium, and thorium. Because the crust conducts heat poorly, this heat remains where it is produced, dissipating only slowly. Near the surface, there is usually a layer of sediments and fractured rock kept cool by groundwater circulating through it. Below this layer, the temperature begins to rise. In stable geological zones (close to 90 percent of the earth's surface), the average rate of temperature increase, or thermal gradient, is about 25°C/km.

In some locations, however, geophysical activity — most often associated with the movement of tectonic plates along fault lines — allows molten rock to approach closer to the earth's surface, thus increasing the thermal gradient and making geothermal energy more accessible. Such hyperthermal regions exist in several parts of the world, including the western United States, east Asia, the Mediterranean, and Iceland. Where surface water seeps down far enough to come into contact with hot rock, its density is reduced, creating sufficient buoyant pressure to bring it back to the surface, where it emerges as hot springs or steam. In some cases, pores and fractures within rock trap this hot water or steam far below the surface, creating hydrothermal reservoirs.

Strictly speaking, geothermal energy is not a renewable resource. Because of the very low thermal conductivity of rock, it may take thousands of years to replace the heat in a large volume of rock once a significant fraction of it has been withdrawn (Armstead and Tester 1987). In addition, where hydrothermal reservoirs are recharged only slowly from surface water, they can quickly be exhausted if too much steam or hot water is removed or not enough is replaced through injection back

Figure 7.1
The 28 steam power plants at The Geysers in northern California represent about one-third of the world's geothermal-electric capacity. However, a drop in steam pressure caused by overproduction has reduced the power output of these plants by 25 percent since 1987. Source: Geothermal Resources Council.

Table 7.1
Geothermal resources of the United States, in exajoules. The "accessible" resource is the total amount in the ground. The "recoverable" resource is the amount that can be recovered at the wellhead.

	Accessible Resource (EJ)	Recoverable Resource (EJ)	Electricity Production (MW for 30 years)
Hydrothermal[a]			
Identified			
Steam (The Geysers)	76-124	5-14	860-2,400
Water (>150°C)	770-930	180-240	17,700-24,300
Water (90°C-150°C)	590-810	121-231	—
Water (<90°C)	26,000-28,000	83-91	—
Undiscovered			
Water & Steam (>150°C)	2,800-4,900	700-1230	72,000-127,000
Water (90°C-150°C)	3,100-5,200	770-1300	—
Water (<90°C)	7,200	66	—
Geopressured[b]			
Sandstone	17,100	?	?
Shale	153,000	?	?
Hot Dry Rock[c]			
High Quality (>50°C/km)	2 million	?	?
Low Quality (<50°C/km)	9.7 million	?	?
Magma[d]	0.1-1 million	?	?

[a] Sources: USGS (1979); USGS (1983). For high- and medium-temperature water-dominated reservoirs, the recoverable resource is defined to be 25 percent of the accessible resource.
[b] Source: USGS (1979). The geopressured resource includes thermal energy and methane, but not hydraulic energy, located in onshore and offshore reservoirs in the northern Gulf of Mexico region.
[c] Calculated using the method described by Armstead and Tester (1987), based on thermal gradient data provided by Tester and Herzog (1990), and assuming a maximum drilling depth of 10 km and minimum useful rock temperature of 85°C.
[d] Source: USGS (1979). The magma resource is defined as the thermal energy in young igneous-related formations, both identified (lower bound) and unidentified (upper bound), to a depth of 10 km.

into the system, even if the temperature of the rock itself is not significantly affected. This seemingly academic point actually has considerable practical significance, in that one of the chief uncertainties in planning a geothermal project of any type is how much energy can be produced before the reservoir's productivity declines substantially. As we will see, a 25 percent decline in steam output has already been observed at The Geysers, the largest geothermal development in the world.

Even so, on a regional and global basis, the geothermal resource is so large as to be virtually inexhaustible. The *accessible resource base* in the United States, broadly defined as the amount of heat above a minimum useful temperature within drilling distance of the surface, is estimated to be anywhere from 1.2 million EJ to more than 10 million EJ, depending on assumptions of required temperature, practical drilling depths, and efficiency of recovery (NAS 1987, Armstead and Tester 1987). By comparison, proven and unproven coal reserves in the United States are estimated to be 85,000 EJ (EIA 1991b).

Such a gross measure of the geothermal resource, however, provides at best only an extreme upper boundary for discussion of its potential contribution to the U.S. energy supply. Estimating the fraction that could be recovered and used at a competitive price is difficult and requires a careful look at the characteristics of each type of resource as well as the technologies that have been developed or are under development for exploiting them. The four main types of resource are hydrothermal fluids, geopressured brines, hot dry rock, and magma.

Hydrothermal Fluids

Hydrothermal reservoirs are the only type of geothermal resource currently being exploited for commercial energy production. Most of the reservoirs that have been identified were discovered because of the discharge of hot water or steam at the surface, although some have been found accidentally while drilling wells for other purposes or through deliberate exploration. Undoubtedly, many reservoirs remain to be discovered.

Hydrothermal reservoirs are classified as either water-dominated or steam-dominated. The latter are quite rare: Only three have been identified in the United States, one at The Geysers, another (unconfirmed by drilling) near Morgan Springs, California, and the last in Yellowstone National Park, Wyoming.[1] Easily the largest is The Geysers, which the U.S. Geological Survey estimates contains about 100 EJ of thermal energy in rock and steam, or slightly more energy than the United States consumes in a year.[2] An additional 210 EJ of thermal energy are believed stored in steam-dominated reservoirs that have not yet been identified (USGS 1979). Water-dominated reservoirs, which are much more common, are classified by temperature: high (greater than 150°C), intermediate (90-150°C), and low (less than 90°C). More than 50 high-temperature systems and 200 intermediate-temperature systems have been identified, containing collectively an estimated 1,360 to 1,740 EJ of thermal energy, or roughly 20 years worth of U.S. energy supply. Based on statistical extrapolations, the U.S. Geological Survey estimates that 7,260 to 11,840 EJ are contained in identified and undiscovered reservoirs, split almost evenly between high and intermediate temperatures (USGS 1979). In addition, more than 1,000 low-temperature reservoirs containing as much as 28,500 EJ of thermal energy have been identified. Most of this energy is believed stored in immense warm-water aquifers under the central plains between Texas and Montana (USGS 1983). Although low and intermediate temperatures are not suitable for generating electricity, they can be valuable in direct use applications such as space heating.

These estimates are impressive but need to be carefully qualified. For one thing, only a portion of the fluid in a hydrothermal reservoir can be recovered; the rest will remain trapped in rock pores and fissures. The recoverable fraction is commonly estimated to be 25 percent, but may be much lower in some systems. Moreover, only a relatively small fraction of the recoverable heat can be transformed into electricity. Even with high-temperature reservoirs, the efficiency of electricity conversion is quite low — typically 10-15 percent — compared to efficiencies of around 33 percent achieved in conventional steam-electric power plants. To put the resource in more practical terms, it is estimated that high-temperature reservoirs in the United States could sustain 90,000-150,000 MW of power production — equivalent in annual generation to 30-45 percent of current U.S. electricity needs — for 30 years.

Geographic constraints are also an important consideration. All of the known high-temperature hydrothermal reservoirs, whether water- or steam-dominated, are located in western states. Lower temperature reservoirs are more widely distributed, but still cluster west of the Mississippi River and generally in areas far from population centers. For direct heat applications, in particular, it is crucial that the heat source be located near the demand, as piping costs and energy losses increase rapidly with distance.

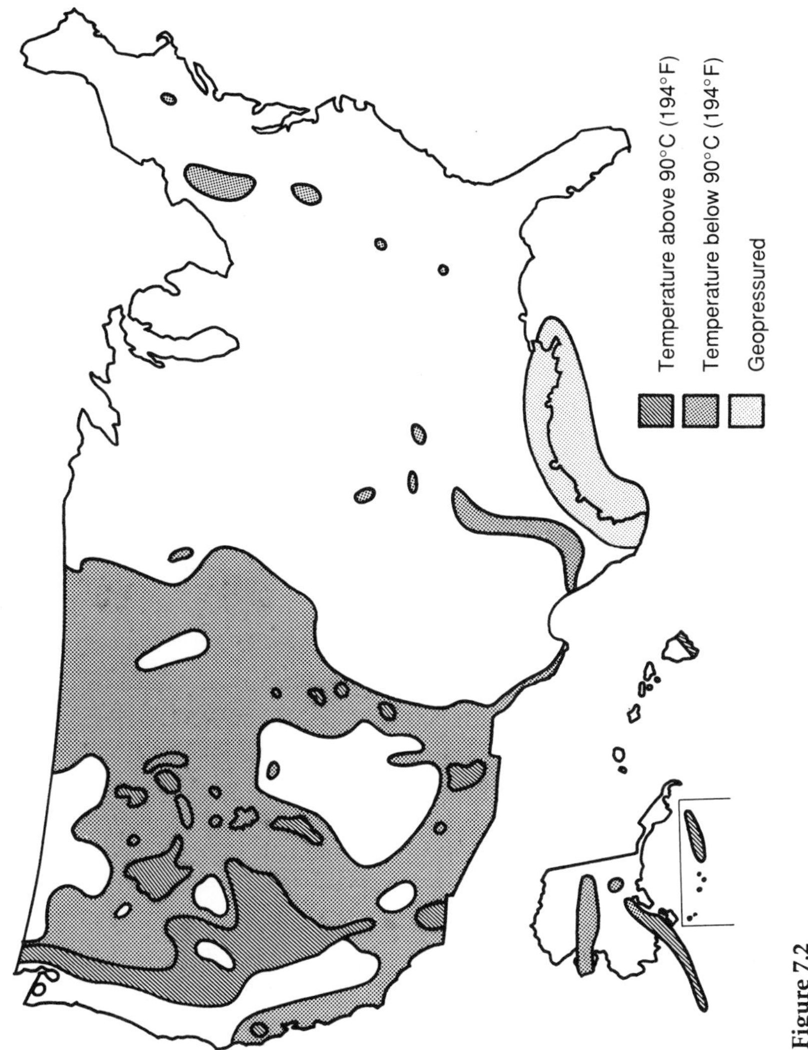

Figure 7.2
Map of U.S. hydrothermal and geopressured resources. Source: U.S. Department of Energy.

Electricity Production

By far the widest current use of hydrothermal energy is electricity production. As of 1990, the world's installed capacity of geothermal power plants was 5,828 MW, of which almost half, 2,770 MW, was located in the United States (Huttrer 1990). U.S. development in the past decade has been astonishingly rapid: Total geothermal-electric capacity more than doubled from 1985 to 1990, largely because of opportunities created by the 1978 Public Utilities Regulatory Policy Act (PURPA, described in chapter 3). California produces the most geothermal energy of any state and gets about 6 percent of its electricity from this resource.

Most of the geothermal power plants in operation today — those built at The Geysers and Larderello, Italy — use dry (superheated) steam as their energy source. It is generally believed that rock fissures in these reservoirs are filled with dry steam or in some sections with water that flashes to steam as it rises and the pressure is reduced. Once extracted at the wellhead, the steam is directed through a turbine to turn a generator. It then returns to liquid in a condenser so as to reduce the pressure at the turbine outlet and improve thermal efficiency. Modern dry steam power plants at The Geysers have a net thermal efficiency of about 15 percent, and the cost of electricity is around 4 or 5¢/kWh, including the cost of the steam, which is purchased from the owners of the steam field (EPRI 1989c).

Most of the world's known and accessible dry steam reservoirs have now been fully exploited, so the trend is toward increased use of water-dominated reservoirs. Several hot-water plants have been built in areas such as Coso Hot Springs, California, Puna, Hawaii, and Steamboat Springs, Nevada. Because of their greater complexity and intrinsically lower efficiency (6 to 12 percent), and because of the need to dispose of large volumes of often contaminated water under environmental restrictions, hot-water plants are usually more expensive to build and operate than dry steam plants.

There are two basic kinds of hot-water power plant, the flash plant and the binary plant. In a flash plant, a portion of the water is flashed to steam by dropping the pressure in a chamber called a separator, in an imitation of the natural steam generation that occurs in The Geysers. Most flash plants have either one or two flashing stages, although a few have three; the second and third stages, which operate at successively lower pressures, provide increased efficiency. In a binary plant, the water does not drive the turbine directly; rather, its heat is transferred in a heat exchanger to another fluid, which in turn is vaporized and directed through the turbine. This method is generally favored where the reservoir temperature is below about 190°C, because then the secondary fluid (usually isobutane or freon) is a more efficient flashing medium than water, or when the reservoir is contaminated with a high concentration of dissolved gases, which can damage turbine blades. Most hot-water power plants in operation today are flash plants, but several binary power stations have been built since the late eighties.

Despite the considerable success of hydrothermal-electric technology in the past two decades, the industry's reputation has been damaged by the sharp and widely unexpected drop in steam pressure observed at The Geysers since 1987. The power plants there were designed for a capacity of about 2,000 MW, but by 1991 their average output had fallen to 1,500 MW. If this trend continues, output may slip to about 1,000 MW by the end of the nineties. This could mean severe financial losses for investors. Plant owners have responded to the problem by "cycling" some plants so they generate power only in peak periods, when utilities are willing to pay more for the electricity. But any way it is viewed, The Geysers experience has been a deep disappointment — and a telling lesson — for the geothermal industry.

The basic problem at The Geysers is that more steam is being withdrawn from the reservoir than is being replaced through injection and natural recharge, or as one developer put it, "there are too many straws in the teapot" (Kerr 1991). In hindsight, independent power producers, utilities, and state regulators moved too quickly to expand capacity at The Geysers in the mid-eighties while failing to heed geologists' warnings that the reservoir's productivity could decline more quickly than expected. More than a decade ago, the U.S. Geological Survey estimated that The Geysers could sustain 1,630 MW of production for 30 years, but the range of uncertainty on this estimate was plus or minus 770 MW (USGS 1979). As it turned out, the lower bound estimate was closer to the truth.

It is probably too late to stop, let alone reverse, declining steam pressure at The Geysers. Injecting cold water into the reservoir will help, but recent experiments with this by steam producers have had mixed results (Lippmann and Bodvarsson 1990). Looking on the bright side of things, however, over 95 percent of the heat originally contained in The Geysers remains there within the reservoir rock. Thus it may be possible to regenerate the reservoir, perhaps using hot dry rock technology (Brown 1990). More than anything, the experience at The Geysers brings home the need for careful, conservative evaluation of hydrothermal reservoirs before serious development begins.

Direct Use

Electricity production has been the dominant, but not by any means the only, application of geothermal energy in the past century. Since the first district heating project in Boise, Idaho, in 1892, there has been a steady, if slow, expansion in the number of projects in the United States that use geothermal heat directly. As of 1990, according to one survey, about 20 quadrillion joules (0.02 EJ) per year of direct heat were being produced from geothermal sources (Lund, Lienau, and Culver 1990).

The various uses of geothermal heat depend on the temperature of the resource at hand. At temperatures below about 50°C, the most important applications are greenhouses and aquaculture. At temperatures of 50-100°C, the market is dominated by hot water and space heating for buildings. At still higher temperatures, industrial applications such as the manufacturing of chemicals, pulp, and paper, as well as food dehydration and canning, predominate.

At the top of the list of current uses is enhanced oil recovery. In these operations, under way in several midwestern states, warm water from thermal wells is pumped into oil wells to help reduce oil viscosity and promote production (Lienau 1991). Second on the list is space heating and cooling, which encompasses a variety of technologies. In some projects, hot water is distributed from a central source through pipes to tens or hundreds of homes and buildings. In others, buildings are heated from individual wells, often by placing a heat exchanger in the well and circulating water through it. Most such applications are very inexpensive. For example, Elko Heat Company in Nevada sells thermal water to its district heating customers for $0.375 per therm, or $3.50/GJ, assuming a 22°C drop in water temperature after use. This is about two-thirds the current average residential price of natural gas.

Hot water can also drive thermal-absorption chillers for cooling buildings, but only a few such systems have been installed because the technology is practical only on a large scale. One example is the Oregon Institute of Technology in Klamath Falls, where several buildings are cooled by cold water piped from a lithium-bromide thermal-absorption chiller (Anderson and Lund 1987).

In recent years, ground-source heat pumps have emerged as a popular technology for heating buildings. Working on the same principle as a refrigerator, a ground-source heat pump uses electricity to transfer heat from the earth to a building or, when operating in the cooling mode, from the building to the earth. Some heat pumps use a buried heat exchanger; others use water from a nearby well or pond. As of 1990, some 80,000 heat pumps were in operation, and the number is growing at a rate of about 20,000 per year (Lund, Lienau, and Culver 1990). By comparison, only 27 direct-heat projects of all types were completed or under construction from 1985 to 1990, although some of these involved hundreds of homes.

The main advantage of ground-source heat pumps is that they require no hydrothermal reservoir and thus can be used anywhere in the country. In addition, because the temperature of the earth below a few meters depth varies much less than the temperature of the air throughout the year, ground-source heat pumps are about 30 percent more efficient, on average, than the more common air-source heat pumps, and are particularly advantageous in cold climates. Of the 80,000 heat pumps in operation, 50,000 use groundwater as their heat source and sink and the rest are earth-coupled. Of the latter, two-thirds use vertical and the rest horizontal earth-coupling loops. The horizontal types, since they are buried less than 10 meters (30 feet) below the surface, depend more on the average annual ambient air temperature than on geothermal heat for their efficient operation (Lund, Lienau, and Culver 1988).

Other commercial applications of direct geothermal heat include greenhouses and aquaculture. By maintaining steady temperatures year-round, geothermal heat can greatly increase yields of both plants and fish at relatively low cost and can permit operation in colder climates than would otherwise be economically feasible. From 1985 to 1990, five hectares of geothermal greenhouses were built in Montana and New Mexico, and four projects for raising catfish, trout, and other fish were completed in Arizona (Lund, Lienau, and Culver 1990).

Only a very few industrial geothermal projects have been developed, in part because industrial processes often require higher temperatures than are available from local hydrothermal sources. So far, food drying and dehydration involving temperatures of 100-150°C have been the most successful in the United States. Recently, mining companies have begun to use geothermal heat to enhance gold recovery in heap leaching operations in Nevada (Trexler, Flynn, and Hendrix 1990).

One important consideration affecting the future development of hydrothermal direct-heat applications is the proximity of known reservoirs to centers of heat demand. Unless a project is very large, it is not economical to pipe heat over a distance of more than a few kilometers, and because western states are relatively unpopulated, the potential market for heat from hydrothermal sources is quite small compared to the total U.S. heat demand. One survey identified 373 cities in eight western states located within eight kilometers of a thermal well or spring having a temperature of at least 10°C (50°F); yet the combined population of those cities was only 6.7 million, and their combined residential heating demand was just 0.14 EJ/year (Eliot Allen & Associates 1980).

Even where suitable local resources exist, direct geothermal use is not being developed as rapidly as its economic advantages might suggest. Part of the problem is that developing a new hydrothermal field is a costly and sometimes risky venture. The field must be identified and mapped, test wells drilled, and chemical analyses performed, and the reservoir must be tested before commercial production can be considered. Moreover, district heat projects — those most likely to find wide application in the future — are economical only on a substantial scale, usually involving many buildings that are relatively densely spaced to minimize piping costs. Electric and gas utilities with the institutional capacity to pursue geothermal district heat projects have little interest in them and may even regard them as competitors to their traditional business.

Geopressured Brines

Scattered around the United States and the world, large, confined reservoirs of hot, geopressured brine can be found at depths of 3,000 meters (10,000 feet) to more than 6,000 meters (20,000 feet). These unique resources, a hybrid of geothermal energy and fossil fuel, exist where the normal outward heat flow in the earth is stopped by insulating, impermeable sand and clay or shale beds in a region of subsiding crust. Water is produced by the compaction and dehydration of marine sediments, and into this water, methane and other hydrocarbon gases become dissolved (GRRC 1972). Geopressured brines consequently contain three potentially useful forms of energy: heat at temperatures ranging from 130°C to 260°C (270°F to 500°F); hydraulic pressure ranging from 170 to 240 bars (2,500 to 3,500 pounds per square inch) at the wellhead; and dissolved natural gas at concentrations of up to 100 standard cubic feet (scf) per barrel of brine (Lunis 1990). Research on more than a dozen test wells over the past 17 years indicates these brines can be tapped with existing technology, and an industrial consortium formed in 1990 has begun investigating their use (Negus-deWys 1990).

The most intensively studied geopressured system in the United States stretches along the northern coast of the Gulf of Mexico in Texas, Louisiana, and Mississippi. Other significant reservoirs are known to exist in central California, the Rocky Mountain region, and elsewhere. The U.S. Geological Survey has estimated that the Gulf Coast resource, considering both onshore and offshore areas, contains about 11,000 EJ of thermal energy and 6,000 EJ of natural gas trapped in sandstone reservoirs, with perhaps ten times that amount trapped in shale, to a depth of 7,000 meters. Excluding the offshore areas, the accessible thermal resource in sandstone deposits is estimated to be about 5,800 EJ and the accessible natural gas resource about 3,200 EJ (USGS 1979). Since sandstone is more porous, it could sustain much higher flow rates than shale and would be developed first.

Just as is the case with hydrothermal reservoirs, the fraction of geopressured brines that might actually be recovered is difficult to determine. The main limitation is that the removal of fluid from a reservoir decreases its internal pressure, and below a certain critical pressure (which differs from well to well) an economic rate of production cannot be sustained. Initial concerns that geopressured-geothermal extraction could cause land subsidence or microseismic activity have so far not been born out by well tests, and researchers now suspect that subsidence will not be a problem because reservoirs are recharged through leaky faults (Negus-deWys 1992). Even assuming that only five percent of the brines in sandstone deposits in onshore areas can be recovered, the practical geopressured resource is about as large as America's proven reserves of oil and gas.

Technology and Cost

The hot brine from a geopressured well can be used in much the same manner as fluids from a hydrothermal reservoir. The main difference is that the brine is under high pressure and usually contains higher concentrations of dissolved minerals and gases, including, of course, methane. High pressures and temperatures inside the well keep the gases in solution, much like carbon dioxide in a capped bottle of soda. Once the well is tapped, hydraulic pressure forces the fluid to the surface, and there the gases bubble out in a separator. After it is used, the bulk of the separated brine is injected into another well at a sufficient depth to avoid contaminating freshwater aquifers.

In most cases, the gas obtained from a well must be filtered to separate methane from impurities such as carbon dioxide, which usually makes up about 10 percent of the gas volume. Recovered methane can be passed directly into a delivery pipeline, compressed and transported, converted to methanol, or burned on the spot for electricity. In addition, heat extracted from the brine can be used either directly for space heating or other purposes or to generate electricity.

The Department of Energy has funded research into geopressured-geothermal technology since 1975. The program's goal is to demonstrate the capacity to produce electricity from geopressured sources at a levelized cost of 7–11¢/kWh by 1995 (DOE 1989). So far, thirteen wells, including nine "wells of opportunity" (that is, abandoned oil and gas wells) and four specially drilled wells, all in Texas and Louisiana, have been studied, with extensive testing on three of these.

The Pleasant Bayou Well, located 50 miles south of Houston, taps into a reservoir with an estimated brine volume of 8 billion barrels. From September 1989 to May 1990, researchers successfully tested a 1 MW hybrid power plant at the well. In this plant, recovered gas was burned in two engines to generate electricity while the waste heat from the engines, combined with the hot brine, was passed through binary heat exchangers to vaporize isobutane and run a separate turbine. Like most such systems, this plant was not designed to capture hydraulic power from the well (Negus-deWys 1990). Although some scaling and corrosion of the heat exchangers occurred during the test because of the high levels of dissolved solids in the brine, researchers believe these problems will be manageable. They have developed a solution for scale control that involves pumping phosphorate inhibitors into the reservoir; these appear to attach to sand grains surrounding the wellbore and are gradually released into brine flowing into the well, thus providing a "time-release" of chemicals that can prevent scaling for up to 13 months. Corrosion can be controlled by proper design and choice of materials (Fortuna and Jelacic 1989).

Another reservoir under study is one tapped by the Willis Hulin Well in Vermilion Parish, Louisiana, a commercial well that produced natural gas for 19 months before pressure dropped and it was transferred to the Department of Energy. The depth and borehole diameter of this well limits production to under 20,000 barrels per day, however, which is not enough to be economic. This performance may be typical of reworked oil and gas wells (Negus-deWys 1990).

The third main exploration well, located in Cameron Parish, Louisiana, produced 27 million barrels of brine and 676 million scf of gas from 1983 to 1987, when it was capped to study how partially depleted geopressured reservoirs recover over time. During its nearly five years of production, wellhead pressure dropped over 40 percent. By late 1991, the pressure had returned almost to preproduction levels (Negus-deWys 1992).

Early research on other potential geopressured reservoirs in south Texas indicates that some may be less than 1,800 meters deep and contain brine as hot as 260°C and saturated with up to 100 scf of natural gas per barrel. Although it is much less expensive to use old oil and gas wells for exploiting geopressured brines — reworking the Hulin Well cost one-quarter as much as drilling a new one — the best reservoirs may lie in areas not yet studied.

In addition to, or in combination with, generating electricity, geopressured brines could also provide direct heat for many applications. They could be used for thermally enhanced oil and gas recovery, an especially attractive option in areas where fresh water is scarce. They could also warm greenhouses and aquaculture operations or circulate under citrus fields, in the latter case extending the growing season and providing protection against frost. The most ambitious schemes would involve a cascade of uses, in which, for example, hot brine would first be used to generate electricity, then circulated through a food processing plant, then, at much lower temperatures, through a greenhouse or aquaculture facility, before finally being injected back into the ground (Lunis 1990).

While research to this point has demonstrated that the basic technology is in hand, researchers need to be able to predict a reservoir's productive lifetime with higher confidence before geopressured brines can be commercially developed. The most likely near-term use of this resource will be for enhanced oil recovery, the most cost effective application identified so far; commercial electricity generation is regarded to be about ten years away.

Hot Dry Rock

The one geothermal resource not limited by geographic considerations is hot dry rock (HDR). As the name suggests, this resource originates from rock that contains few or no natural fluids to transfer the heat to the surface, yet is sufficiently hot to be a potentially useful source of energy.

Figure 7.3
Map of U.S. geothermal temperature gradients, or temperature increase with depth. Areas of high or moderate gradient are more economical for hot dry rock development, but direct heat applications may still be feasible in nonthermal areas. Source: U.S. Department of Energy.

The HDR resource is extremely large, but the technology for exploiting it, which involves fracturing the rock and passing water between two or more wells, is complex and still under development.

Hot dry rock is found at some depth everywhere around the globe. The main requirement is that the rock be sufficiently rigid, or competent, to be easily fractured so water can be passed through it; this usually means granite, limestone, or basalt. Alternatively, naturally fractured rock can sometimes be used. The economic feasibility of exploiting this resource, however, depends strongly on the local thermal gradient, or rate of temperature increase with depth. A high gradient means that a relatively shallow well can reach rock of a desirable temperature. In the low-grade, or nonthermal regions, typical of the eastern United States, the thermal gradient averages about 25°C/km, whereas in the high-grade, or hyperthermal, regions typical of the western United States, it often exceeds 50°C/km.

Since high-grade areas are likely to be developed first, it is useful to estimate this resource separately.[3] According to the most recent data, thermal gradients exceeding 50°C/km are found over about 5.5 percent of U.S. land area (Tester and Herzog 1990). Assuming an average thermal gradient of 60°C/km, a maximum drilling depth of 10 km, and a minimum useful temperature of 85°C,[4] the amount of usable thermal energy stored in HDR in those areas is about 2 million EJ. Taking 25°C/km as the average thermal gradient in nonthermal regions, then the low-grade resource is about 9.7 million EJ, or almost five times as large. The total combined accessible resource is about 11.7 million EJ, or enough, in principle, to meet current U.S. energy needs for 130,000 years.

Different assumptions in this calculation will yield different results. For example, if the minimum useful temperature is defined to be 125°C, rather than 85°C, then the total resource is reduced by 40 percent to about 7.1 million EJ. If, in addition, the maximum drilling depth is defined to be 6 km — judged the current practical limit of HDR technology — then the total resource is reduced by an additional 86 percent to 990,000 EJ. Even this restricted resource, however, could theoretically supply current U.S. energy needs for more than 10,000 years.

Technology and Cost

How much of this immense resource could actually be tapped depends on the feasibility and cost of HDR technology. Since 1970, several countries, including the United States, Great Britain, Germany, Japan,

France, Sweden, Switzerland, and the Soviet Union, have conducted HDR research programs. The most advanced is the Fenton Hill project in the United States, directed by researchers at the Los Alamos National Laboratory, but important work has also been done at the recently terminated Rosemanowes project in Great Britain and, since the mid-eighties, in Japan. These efforts have resolved many of the crucial questions originally raised concerning this technology, leaving at least some researchers confident that with a modest amount of additional development, HDR could become commercially competitive with fossil fuels for both electricity generation and direct heat applications by the turn of the century (Armstead and Tester 1987, Duchane 1991).

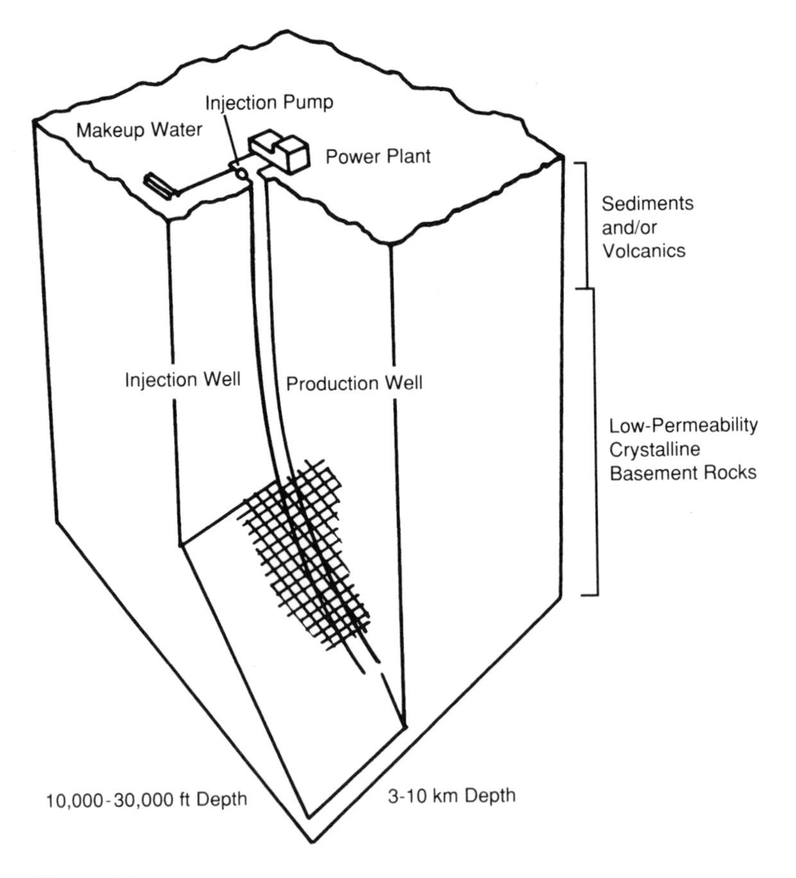

Figure 7.4
Artist's rendering of the hot dry rock concept. Source: U.S. Department of Energy.

Scientists at Los Alamos National Laboratory were the first to propose, in 1970, that heat could be extracted from impermeable rock by drilling two almost adjacent wells, fracturing the rock between them using hydraulic pressure (a technique originally developed for oil recovery), then pumping water down one well and out the other. Flowing through the cracks in the rock, the water would absorb heat along the way and come out much hotter than it went in. In essence, this method creates a giant heat exchanger, or equivalently, an artificial hydrothermal reservoir. This ambitious scheme was first tested in 1977 at Fenton Hill, New Mexico, located a short distance from Los Alamos on the western side of the Valles Caldera, a large, extinct volcanic formation. A key advantage of this site is its high thermal gradient, which averages about 60°C/km.

In Phase I of the Fenton Hill project, a well was drilled to a depth of just under 3 km, vertical fractures were created in the rock around it by pumping in fluids at high pressure, and a second well was drilled at a slightly shallower depth to intersect the fractures. Preliminary flow tests showed insufficient surface contact between the fractured rock and water pumped through it, so the fracture zone was enlarged. By the end of subsequent testing in 1980, the system was producing pressurized hot water at about 135°C. Some of this was diverted to run a 60 kW power plant supplying electricity for the site — the first time power had ever been produced from an artificial geothermal system.

Table 7.2
Results of an economic assessment of the cost of electricity from hot dry rock for typical sites in the West, Midwest, and East, and for today's technology, initial commercial technology, and "optimized" technology.

Region (Gradient)	Cost[a] (1990 ¢/kWh)		
	Today's Technology	Commercial Technology	Potential
West (80°C/km)	5.7	4.3	2.9
Midwest (50°C/km)	9.3	6.5	3.8
East (30°C/km)	28.9	17.3	8.5

[a]Adapted from Tester and Herzog (1990). Data were converted to 1990 constant levelized cents per kilowatt-hour using a 12.3 percent fixed charge rate.

Phase II of the project was designed to approximate more closely the operating conditions of a commercial power plant. It involved drilling two new wells and creating a larger fracture zone at a depth of 3.6 km, where the temperature is 240°C. The scientists experienced a setback when they discovered that the fractures did not connect the two wells, but solved the problem by using microseismic (acoustic) signals to map the fracture zone, then redrilling part of the upper well to intersect it. By the end of preliminary flow tests in 1986, the system was producing 235 gallons of pressurized water per minute at a temperature of 190°C (375°F), corresponding to a thermal production rate of about 30 GJ per hour. Further flow testing was delayed by repairs to a well casing. After tests to verify this well's integrity and the construction of a permanent test facility on the surface, researchers began an extended flow test program in late 1991 to evaluate the new reservoir (Duchane 1992).

The results obtained from Fenton Hill and HDR projects elsewhere have provided strong evidence of the technical feasibility of the hot dry rock concept. To make commercial development possible at anything other than the very best sites, however, at least three requirements must be met. First, drilling costs will have to be reduced, as they represent half or more of the total investment cost of HDR projects. Drilling through hot granite creates severe thermal and physical stresses on conventional rotary-drilling components. Advanced techniques such as thermal spallation (using a hot gas jet to break up rock) and erosive drilling are being studied, but they have not been adequately demonstrated in the field.

Second, improved fracturing techniques are needed to enhance the efficiency of heat exchange in the fracture zone and thereby raise the productivity of the HDR reservoir. Ideally, a reservoir should combine several qualities, including high initial temperature, large fracture surface area, large connected volume, low resistance to water flow, and minimal water losses into surrounding rock. No single experimental reservoir has succeeded in meeting all of these requirements. For example, flow resistance in the Fenton Hill Phase II reservoir remains disappointingly high, while excessive water losses have plagued both the British project at Rosemanowes, Cornwall, and the Japanese HDR experiments, finally causing the former to be canceled in early 1991.

Third, researchers will have to demonstrate sufficient understanding of the behavior and characteristics of HDR reservoirs to establish a credible basis for predicting their useful lifetime. Like all geothermal reservoirs, HDR reservoirs will be depleted over time as their stored heat is extracted. In commercial systems, the thermal drawdown rate must be kept within reasonable limits — perhaps one or two percent per year. Otherwise, after only a few years, costly new wells will have to be drilled or the reservoir refractured to restore fluid temperatures to their original level. Predicting reservoir performance from models and flow tests remains more art than science.

Assuming that these hurdles can be overcome — still an open question — analysis suggests that HDR technology could provide electricity and heat at a relatively low cost, at least in areas of moderate and high thermal gradient. For example, calculations by Tester and Herzog (1990) indicate that even with today's immature technology, electricity produced at high-grade HDR sites would be strongly competitive with that from conventional baseload sources. With mature technology, it should be possible to produce competitively priced electricity even at medium-grade sites, thus opening up much of the midwestern and southern United States for development. Low-grade sites in the East will probably not be economic for electricity generation under any conditions. Results such as these have led to proposals for building the first commercial HDR plant at Roosevelt Springs, Utah, or at Clear Lake, California, both areas with extremely good HDR resources.

Analysis also indicates that even low-grade areas could be developed for commercial direct-heat applications because in this case it would not be necessary to drill very deep — perhaps 3,000 meters — to reach rock of a useful temperature. Thus, by drilling HDR wells near centers of heat demand, it should be possible to avoid the geographic limitations of hydrothermal and geopressured resources and make geothermal heat available at a reasonable cost throughout the country.

None of this is likely to happen soon, however, unless more federal funding is put into the development and demonstration of hot dry rock technology. The program is currently underfunded; researchers are not even assured of having enough money to complete the planned one- or two-year extended flow test at Fenton Hill, let alone construct a commercial-scale demonstration facility, which is the next logical step in the program. The California Energy Commission is funding a detailed assessment of the HDR resource at Clear Lake, but no company is likely to take the risk of building the first power plant there without significant federal or state support. American leadership in this field is in danger of being lost in the next few years: Japan's HDR program, which runs experiments at three different sites, is funded at twice the level of the

American program and has already developed some innovative ideas, such as the concept of creating several HDR reservoirs at different depths in a single well to minimize drilling costs.

Magma

If hot dry rock represents a potentially vast source of energy, then magma — molten or partially molten rock — is "an even more tempting prize" (Armstead and Tester 1987). In most continental areas, magma lies at least 35 kilometers below the earth's surface, making it inaccessible to modern drilling techniques. But in areas of current or ancient volcanic activity and tectonic plate movement, magma chambers or protrusions can be found much closer to the surface. With temperatures ranging from 600°C to 1,300°C, magma represents by far the most concentrated form of geothermal energy. As a rough rule of thumb, the thermal energy in 2 cubic kilometers of magma could run a 1,000 MW power plant for 30 years (Eichelberger and Dunn 1990).

Theoretically, great quantities of energy could be produced by tapping magma. Estimates of the resource base are highly uncertain, however, as very little is known about the location, size, and temperature of magma concentrations within reach of the earth's surface. According to the U.S. Geological Survey, at least 101,000 EJ, and perhaps more than 1 million EJ of magma energy might be accessible at depths of less than 10 km in the United States (USGS 1979). Like hydrothermal reservoirs, however, most magma sites would be located in western states, probably making them less useful than hot dry rock.

Subsurface magma chambers are difficult to detect. They can usually be found directly beneath surface volcanic vents, although if the crust has moved they may sometimes be located several kilometers away. (The Hawaiian Islands were formed by the movement of the crust over a deep magma source, which erupted periodically.) Geophysical and geological analysis can point to where likely magma deposits might be found, but cannot establish if they are molten or solidified. Several promising seismic techniques are now being tested; proving their effectiveness, however, will require matching data with drilling samples (Eichelberger and Dunn 1990). Sizeable reservoirs of molten magma are known to exist in Alaska, Idaho, California, and other western states. One of the largest, with an estimated heat content of 36,000 EJ, may be under the Yellowstone Caldera in Wyoming.

In addition to thermal energy, magma also contains potential chemical energy. Water injected directly into magma that is high in ferrous oxide could yield significant amounts of hydrogen gas. Injecting a water-biomass mixture could produce methane and carbon dioxide, or, at higher temperatures, carbon monoxide and hydrogen (Armstead and Tester 1987).

Despite popular images of red-hot volcanic flows destroying everything in their path, drilling into magma is not as far-fetched as it might seem. In 1959, the Kilauea Volcano in Hawaii's National Volcano Park erupted, leaving a modest-sized lava lake in its crater eventually covered over with crust. Using insulated drill pipe originally designed for thermally enhanced oil recovery, researchers in 1982 drilled a well more than 100 meters into the magma. High velocity water jets preceding the drill bit solidified the magma in its path, creating a stable borehole and preventing magma from flowing back up the well. This test showed that if the drill pipe is constantly insulated with relatively cool circulating fluids, its temperature can be kept below 230°C even while deep within 1,000°C magma (Chu et al. 1990).

From this well and subsequent laboratory work, designs for a working magma energy extraction operation have emerged. Drilling from the surface to the edge of the magma chamber should be relatively easy. Insulated, concentric pipe would be placed to the edge of the magma chamber; only the center pipe, however, would continue into it. Once the molten rock was penetrated, water or another working fluid would be kept in constant circulation to cool the drill bits and pipe as well as solidify the surrounding magma.

After the drilling was complete — initial plans call for 1 kilometer penetration into magma — this solidified rock would be fractured thermally or hydraulically to create a large zone for heat exchange. Water pumped to the bottom of this zone would extract heat from the fractured rock as it percolated upward. This hot water would eventually funnel toward the outer pipe at the magma margin and return to the surface. Heat removal from the fractured rock would set up convection currents in the surrounding magma, replenishing thermal energy inside the well.

At the surface, a binary power cycle would probably be the best for generating electricity. Power output would be determined primarily by the flow rate of the working fluid. Although faster flow rates might initially provide increased power output, over the long run the result would be to decrease the well's operating temperature by overcooling and thickening the fractured zone, thus reducing heat transfer between molten magma and the wellbore. A flow rate of about 50 kg/second is estimated to provide an optimal rate of energy extraction, implying a net power output of 25 to 45 MW (Chu et al. 1990).

Magma researchers face major hurdles in commercializing their technology, however. Constant cooling of the drill pipe and fracture zone must be maintained at all times. Even a temporary loss-of-cooling accident could cause solidified magma to liquify, essentially melting away the bore hole and possibly the drill pipe. Another area of concern is corrosion. Magma contains many reactive chemical species, and the high well temperatures would speed corrosion of drill bits, pipe and other equipment. Tests show that carbon steel would be quickly destroyed, while chromium alloys would be more resistant to the oxidizing environment. In hydrothermal brines, however, the latter reveal severe pitting and cracking.

Some of these problems will be explored at the Long Valley Caldera in California. This depression, formed by the collapse after a volcanic eruption 700,000 years ago, sits about 50 kilometers southeast of Yosemite National Park atop what is believed to be a still-molten magma chamber. The centerpiece of the Department of Energy's magma research program, this 6,000 meter (20,000 foot) well is expected to confirm the presence of magma as well as test high-heat drilling techniques. The first phase of drilling to 750 meters (2,500 feet) was completed in 1989, after which the project was suspended when federal funding for magma research shifted toward projects with more immediate benefit for industry. The California Energy Commission, however, has allocated funds for additional drilling at the site, and is including magma in its plan for energy in the 21st century. In 1991, the well was further drilled to 2,000 meters (6,800 feet), and current plans call for its completion in 1995. Even if tests show the presence of magma at Long Valley, commercial power production is still at least 20 years in the future (Finger 1991).

Environmental and Siting Issues

While the four geothermal resource types discussed above differ in many respects, they raise a similar set of environmental issues. Air and water pollution are two leading concerns, along with the safe disposal of hazardous waste, siting, and land subsidence. Since these resources would be exploited in a highly centralized fashion, reducing their environmental impacts to an acceptable level should be relatively easy.

Siting new plants in scenic or otherwise environmentally sensitive areas will always be difficult, however (NAS 1987).

The method used to convert geothermal steam or hot water to electricity directly affects the amount of waste generated. Closed-loop binary systems are almost totally benign, since gases or fluids removed from the well are not exposed to the atmosphere and are usually injected immediately after giving up their heat. While this technology is more expensive, in some cases it may reduce scrubber and solid waste disposal costs enough to provide a significant economic advantage.

Open-cycle systems, on the other hand, may generate large amounts of solid wastes as well as noxious fumes that must be controlled. Most of the pollution and wastes produced are gases, minerals, and metals dissolved in geothermal steam and hot water. Chemical reactions occur much faster at high temperatures, so geothermal fluids easily leach substances from the rocks they pass through. At hot springs, fumaroles, and natural geysers like Old Faithful, these are spewed into the atmosphere, though usually in amounts too small to have much local impact. When geothermal fields are tapped for commercial production, however, the large amounts of chemicals released can be hazardous or objectionable to people living and working nearby.

At The Geysers, for example, steam vented at the surface contains hydrogen sulfide (H_2S) — accounting for the area's "rotten egg" smell — as well as ammonia, methane, and carbon dioxide. In 1974, more than 24 tons of H_2S were being produced daily from the power plants, which then had a capacity of 631 MW. After many complaints, plant owners began installing efficient scrubbers on their plants, which so reduced the emissions that by 1986, with 1,800 MW on line, only 4.5 tons were being released daily (Mock and Beeland 1988). Environmentalists generally acknowledge that the problem is under control. Carbon dioxide is also released at The Geysers and other hydrothermal plants and is expected to make up about 10 percent of the gases trapped in geopressured brines. For each kilowatt-hour of electricity generated, however, the amount of carbon dioxide emitted will still be a small fraction — typically 5 percent — of that emitted by a coal- or oil-fired power plant (DiPippo 1990).

Scrubbers reduce air emissions but produce a watery sludge high in sulfur and vanadium, a heavy metal used as a scrubber catalyst that can be toxic in high concentrations. Additional sludge is generated when hydrothermal steam is condensed, causing the dissolved solids to precipitate out. This sludge is generally high in silica compounds, chlorides, arsenic, mercury, nickel, and other toxic heavy metals (Pasqualetti and

Dellinger 1989). Such precipitation is most severe at geothermal plants like those in California's Salton Sea that work with hypersaline brines and generate more than 50 tons of sludge a day. Because of the high concentrations of heavy metals, the sludge must be treated as hazardous waste. One costly method of waste disposal involves drying it as thoroughly as possible and shipping it to licensed hazardous waste sites. Research underway at Brookhaven National Laboratory in New York points to the possibility of treating these wastes with microbes designed to concentrate and recover commercially valuable metals and render the waste nontoxic at the same time (Premuzic, Lin, and Kang 1990).

Usually the best disposal method is to inject liquid wastes or redissolved solids back into a porous strata of a geothermal well. It is crucial to inject the wastes well below freshwater aquifers and to make certain there is no communication between the usable water and waste-water strata. Leaks in the well casing at shallow depths must also be prevented.

In addition to providing safe waste disposal, injection may also help prevent land subsidence. At Wairakei, New Zealand, where wastes and condensates were not injected for many years, one area has sunk 7.5 meters since 1958. No land subsidence has been detected at The Geysers and most other hydrothermal plants in long-term operation, however. Since geopressured brines primarily occur along the Gulf of Mexico coast, where natural land subsidence is already a problem, even slight settling could have major implications for flood control and hurricane damage. So far, however, no settling has been detected at any of the three experimental wells under study.

Unless a geothermal power plant is equipped with air cooling — which only works well at higher altitudes — a large amount of cooling water will be needed for its operation. (In rare cases, like The Geysers field, steam condensate can be used for cooling.) An HDR power plant will require additional water to make up for losses from the production zone into surrounding rock. These demands can raise conflicts over water resources.

Probably the most serious obstacle confronting the development of hydrothermal reservoirs is that they tend to be located in or near wilderness areas of great natural beauty — Yellowstone and Crater Lake National Parks and the Cascade Mountains, to name a few. If hydrothermal-electric development is to expand much further in the United States, reasonable compromises will have to be reached between environmental groups and industry.

The issue is highlighted in Hawaii, where a bitter battle has erupted over a proposed 500 MW project at Puna on the main island. The key issue in this battle is the location of the hydrothermal field in the midst of the Wao Kele o Puna, the largest undisturbed rainforest remaining in the United States. To native Hawaiians, these 27,000 acres — the name means green forest of Puna — are sacred space, home to the goddess Pele, who is said to live in the volcanic caldera below it. The natives are joined by environmental activists who argue that the last large stand of Hawaii's dwindling rainforest, home to hundreds of unique plant and animal species, should be protected.

State officials and developers argue that this project, and the deep-water cable that would link it with the other islands, will reduce the state's costs of electricity, much of which is now generated from imported oil. They also claim that impacts on the rainforest will be negligible, and estimate that only 300 or 400 acres (less than 2 percent) will be affected by drilling operations and plant equipment. Opponents counter that energy efficiency improvements in the state could easily make up the 500 MW loss in electricity generation resulting from canceling the project and point to dwindling production at The Geysers as a reason for caution. Perhaps even more important, they claim that the network of roads and pipes linking the wells will carve up the forest and provide ready avenues for invading plant and animal species that could profoundly change the character of the forest. They also suggest that noise from the project combined with increased hydrogen-sulfide emissions — which forced the state to close temporarily a demonstration hydrothermal project — could drive away the birds that pollinate the ohia trees that dominate the forest.

The issue is a knotty one for both state officials and environmental activists, all of whom would like to reduce fossil-fuel use in the state. Though the project was first announced in 1981, only a few acres have been cleared for test wells. In June of 1991, a U.S. district judge halted all federal involvement in the project until a formal environmental impact statement is issued. Environmental groups are pursuing similar action at the state level, which, if successful, would halt geothermal development in the rainforest for two years or more.

Notes

1. The latter two are in national park areas and hence excluded from development. Reservoirs in national parks are not counted in resource estimates cited elsewhere in this chapter.

2. High- and intermediate-temperature thermal resources are measured with respect to a reference temperature of 15°C. Low-temperature resources are measured with respect to a reference temperature of 25°C.

3. The following calculations are based on the method described by Armstead and Tester (1987).

4. Ten kilometers is the maximum depth that can presently be reached with vertical (nondirectional) drilling techniques, although for the near term, 6 kilometers is probably the practical limit for HDR wells because of limits on borehole diameter at greater depths. A minimum output temperature of 60°C is required for space heating, the largest low-temperature heat market, but owing to inevitable losses in the rock, surface heat exchangers, and piping, a minimum useful rock temperature is probably about 85°C (Armstead and Tester 1987).

8 Energy Storage

For centuries, builders in Mexico and the southwestern United States have used adobe for the simplest form of energy storage — absorbing heat by day and radiating it when temperatures drop at night. Passive solar buildings today are designed with the same idea in mind, and use materials that readily absorb and hold heat. Such a strategy is necessary when energy production and use do not perfectly coincide.

Conventional wisdom holds that inherently fluctuating renewable energy sources such as solar and wind power cannot make significant inroads into the world's energy supply without large amounts of costly storage, but the reality is more complicated, and less gloomy, than this suggests. A variety of storage technologies is already widely used in residential, commercial, and industrial applications. For example, most homes have a hot water storage tank in the basement. More and more office buildings and factories are making ice at night when electricity is inexpensive, then passing air over it during the day for cooling. Utilities are also storing off-peak power, generated by fossil or nuclear fuel, for use during peak periods. All of these applications actually enhance the value and convenience of conventional fuels. In many cases, the use of solar energy requires only minor and inexpensive adjustments in storage, such as installing a larger and better insulated storage tank for a solar water heater. In addition, hydropower and pumped hydroelectric storage, both already in wide use, can be an ideal complement to wind and solar power.

In many solar applications in the near term, fossil fuels like natural gas would be used as a backup, thus reducing or eliminating the need for storage to ensure a reliable energy supply. Residential solar water heating is usually backed up by gas or electricity, for little extra cost. On a much larger scale, solar-thermal facilities such as the LUZ power plants and Industrial Solar heating facilities also have gas backup. In the future, nonintermittent renewable energy sources such as biomass, geothermal, and ocean thermal energy could take over this important function. In fact, it is quite unlikely that wind and solar energy will ever be used widely on their own. Rather, they will be integrated with other renewable energy sources to maximize reliability and flexibility. For example, solar heat could be used to boost the temperature of geothermal sources for electricity production, while biogas could replace natural gas as the backup power source in solar-thermal applications.

This is not to say that storage will not play a critical role in renewable energy's future, but without more detailed analysis than has been done so far it is difficult to predict precisely what its role will be. Clearly, even the smallest remote power system that depends solely on solar or wind power will need storage batteries (unless constant power output is not a requirement), which will substantially increase the total system cost. The larger question concerns electric utility planning, where reliability is of utmost importance. Electric utilities are accustomed to handling the shifting loads of their customers, and thus adding a moderate fraction of variable wind and solar power — say, up to 10 or 20 percent of the total peak system load — will not greatly affect system reliability. Beyond this point, however, some form of electricity storage will most likely be needed, with the amount and cost depending on the detailed match between wind and solar supply and electricity demand.[1] Likewise, if solar heat supplies more than 30-50 percent of a building's or factory's heat load, then thermal storage will also be needed.

With few exceptions, however, storage technologies already exist to handle these varied requirements. While important technical advances must still be made in some areas, all in all, storage does not appear to present an insurmountable obstacle to a full-scale renewable energy economy. This chapter reviews some of the many storage options that are available or under development. These are divided into two categories: thermal energy storage, including sensible, latent, and thermochemical approaches; and electrical energy storage, including batteries, compressed air, and superconducting magnetic systems. The third storage option considered is hydrogen, which can be used for both thermal and electrical applications and therefore stands in a category of its own.

Thermal Energy Storage

Thermal energy storage is especially well suited for solar applications, since sunlight is easily converted into heat. Passive solar construction, described in chapter 3, makes use of what is called sensible heat storage, in which the temperature of the storage medium varies with the amount of energy stored. Two other types are also being investigated: latent heat storage, which makes use of the energy stored when a substance changes from one phase to another (as from ice to water); and thermochemical storage, which uses the energy stored in reversible chemical reactions.

Sensible Heat Storage

Most sensible heat storage systems employ everyday materials such as water, oil, and rock. While these are inexpensive, their low specific heat (energy stored per unit weight) means that large volumes are needed to absorb significant amounts of energy. Sensible heat storage can be used both to improve the efficiency of energy use and to reduce peak electrical loads. At a fitness center in Missoula, Montana, for example, waste heat from exercisers, lighting, and machinery is collected through the building's air conditioning system and stored in an indoor swimming pool. During the day, the pool warms to about 30°C (85°F). At night, it is allowed to cool to 27°C (80°F), thus giving up about 6.9 GJ to heat the building (Bergoust 1990). Electric furnaces with crushed rock storage reservoirs perform a different function: They shift electrical loads from day to night, thus reducing the cost of the electricity consumed. One study showed that in homes where they were tested, consumption of peaking electricity dropped from 50 percent to just 15 percent of total electricity use (EPRI 1989b).

Storage media such as water and crushed rock work just as well for storing solar heat. Another option is the salt pond. In this approach, dense, highly saline water sits on the bottom of the pond, with less dense, less saline layers stratified above. Solar energy penetrates these upper layers and warms the lower layer. Pipes carrying a heat-transfer fluid through the bottom layer extract the heat. A solar pond equipped with a 5 MW power plant has been operating in Israel since 1983, maintaining stable thermal layer temperatures of over 80°C. Smaller, 6 to 100 kilowatt pond systems have been tried in California and Texas.

At very high temperatures, molten salts can provide efficient sensible heat storage. A mixture of 60 percent sodium nitrate and 40 percent potassium nitrate has been tested as both a heat transfer fluid and storage medium at the Department of Energy's Central Receiver Test Facility in Albuquerque, New Mexico (EPRI 1989a), and will be used in the refurbished 10 MW Solar Two facility (see chapter 3). Sunlight heats the molten salts to high temperatures, and the mixture is stored in large insulated tanks for subsequent electricity generation. This approach allows power output to respond to user demand, rather than solar input. Similar mixtures are being tested for storing waste heat generated by industrial processes such as paper and pulp production, iron, steel, and aluminum smelting, and brickmaking (Spanner and Wilfert 1989).

Latent Heat Storage

Latent heat storage is potentially a more efficient and cost effective approach than sensible heat storage. Melting one kilogram of water without changing its temperature, for example, stores 80 times as much energy as does raising its temperature by one degree centigrade. This means that a much smaller weight and volume of material is needed to store a certain amount of energy. In addition, phase-change materials help regulate temperature, an especially useful characteristic for passive solar building design.

One increasingly common application of latent heat storage is in large commercial buildings with sealed windows, which often require year-round cooling to remove the heat generated by people, computers, lighting, and machinery. Commercial cooling can account for up to 40 percent of a utility's peak demand on a summer day. Prompted by reduced costs for off-peak electricity, some buildings now make ice at night and circulate air over it during the day to cool the building. The Merchandise Mart in Chicago operates the largest such cooling system, which uses more than 900 tons of ice daily (Wendland 1990, Tomlinson and Kannberg 1990).

A variety of phase-change materials with high latent heat capacity are under investigation for storing solar energy. Incorporating paraffins or eutectic salts—easily melted salts with phase-change temperatures near room temperature — into gypsum wallboard for enhanced energy storage capacity is the goal of one Department of Energy program. Such materials must be encapsulated to keep them bound to the board in the liquid phase. High-temperature phase-change materials such as a ceramic-encapsulated mixture of sodium carbonate, barium carbonate, and magnesium oxide, or a silicon-encapsulated alloy of aluminum and silicon, have also been tested for industrial heat storage. Researchers at LUZ International were testing encapsulated phase-change storage materials at one of its 80 MW solar-thermal facilities in California before bankruptcy halted their studies. Such storage would allow the solar-thermal power plants to reduce their natural gas consumption and shift their power output to match more precisely Southern California Edison's electricity loads.

Thermochemical Heat Storage

Reversible thermochemical reactions make up the third group of thermal storage options. Researchers at the Weizmann Institute in Rehovot, Israel, are using the intense heat generated by a solar central receiver to combine methane and steam into carbon monoxide and hydrogen (syngas), which can be stored on-site or piped elsewhere (Garfinkel 1990, Dostrovsky 1991). Exposing syngas to a nickel catalyst causes an explosive reaction which releases heat and regenerates both methane and steam. (Similar reactions are used to produce syngas and methanol from biomass; see chapter 5.) Other possibilities include splitting ammonia into nitrogen and hydrogen, reducing sulfur trioxide into sulfur dioxide and oxygen, or combining calcium oxide with water to form calcium hydroxide. One drawback to all of these systems is their relatively low heat of reaction compared to that of phase-change materials, meaning that relatively large volumes of storage material would be needed for commercial-size systems (McLarnon and Cairns 1989).

Seasonal Heat Storage

The thermal storage systems described above could store heat for several hours or days at most. Seasonal storage is an intriguing concept that could help correct for the mismatch between summer insolation and winter heating demand, which is a major drawback to the use of active solar collection systems for space heating. Seasonal storage absorbs heat generated by solar collectors during the summer and reproduces it for winter heating. This maximizes the efficiency of the collectors, since they can operate throughout the summer when heating demands are low and high ambient temperatures mean lower energy losses from the collectors and piping. This technology is being developed mainly in Sweden, although it is also receiving attention in the United States, which has a more favorable climate for it (Bankston and Breger n.d., Sunderland and Breger 1990).

Isolated subterranean aquifers, excavated earth and clay pits, rock caverns, or ducts drilled in solid rock can all be used for seasonal heat storage. To be both efficient and cost effective, the storage volume must be quite large, implying that the system must serve a large heating load, such as a town or factory. A report from the International Energy Agency (1989), which has sponsored research in this area since 1978, estimates that seasonal storage sufficient to heat around 2,000 homes (180,000 GJ per year) can compete economically with fossil fuels. The large power load and piping costs will limit this type of storage to dense, multihome developments, large buildings, office parks, and shopping centers. In most cases, the area above the storage system can be used for the solar collector field or for the buildings themselves.

Despite continuing research in the United States since the sixties, the first seasonal storage project is only now getting under way here. In a project planned for the University of Massachusetts at Amherst, solar heat stored in a 75,000-cubic-meter clay reservoir will be used to heat a 10,000-seat sports arena and a gymnasium. The preliminary design calls for a flat-plate collector array of 11,000-m^2 surface area to be installed on a 3-hectare (7-acre) field. Beneath this field there is a 30-meter-deep deposit of soft, water-saturated clay with high heat capacity and low groundwater permeability. Up to 600 holes will be drilled into the clay and filled with polymer tubes, then the deposit will be capped with insulation and topsoil (Breger, Sunderland, and Elhasnaoui 1991, Sunderland 1991).

During the summer, hot fluid from the solar collectors will be pumped through the boreholes, warming the clay. During the heating season, water circulated through the borehole heat exchangers will be heated to 55°C (130°F) and piped about 500 meters to the arena and gymnasium. After that, the still-warm water returning to the heat exchangers may be used to melt snow off the large plaza in front of the two buildings. Designers anticipate that the seasonal storage reservoir will be able to deliver about 12,600 GJ annually, providing up to 90 percent of the two buildings' heating and hot water needs. At an estimated initial cost of $3.2 million, the annual constant-dollar solar cost works out to about $11/GJ (assuming an inflation-adjusted financing cost of 4 percent per year). This is about twice the present price of natural gas for residential and commercial users, but since this is a pilot project, one can expect that mature seasonal storage systems will be at least one-third less expensive.

Electricity Storage

Batteries

Batteries, which have been in use for almost 200 years, come in a great many sizes and shapes. Miniature batteries run pacemakers, cameras, and portable radios, while their industrial cousins power forklifts and electric-utility substations. Batteries are already finding wide use in wind and photovoltaic systems for such remote applications as weather monitoring stations, emergency roadside call boxes, marine navigation aids, tower beacons, irrigation water pumping, and vacation homes. Much of the current research in large-scale battery technology is driven by the interest of the electric power industry in shifting loads from peak to off-peak periods.

Another important driving force behind battery development is the looming demand for electric vehicles. New clean-air laws mandating low-emission vehicles will help ensure a large market for electric cars in

Figure 8.1
The GM Impact electric car is the prototype of a vehicle that may be marketed in the mid-1990s. It uses commercially available lead-acid batteries and is projected to have an equivalent gasoline mileage of nearly 70 miles per gallon.

perhaps a dozen states by the turn of the century. California law requires that zero-emission cars make up 10 percent of new cars sold by 2003. In response to this market, leading auto makers in the United States have joined forces with the Department of Energy and the Electric Power Research Institute in a four-year, $260 million project to develop a new generation of batteries for electric vehicles (U.S. Advanced Battery Consortium 1991). This is a vital task, since today's electric cars still suffer significant range and speed limitations, both a result of limitations in battery technology. For example, a Ford Motor Company electric van can travel only about 100 miles before being recharged. General Motor's Impact, a sleek-looking electric car planned for the mid-nineties market, has about the same range and moreover requires three hours for recharging (Wald 1991, Gordon 1991).

All storage batteries work according to the same general principles. Electrodes that are immersed in an electrolyte solution act either as electron donors or receptors. During charging, electricity is converted into the chemical energy of molecular bonds within the electrolyte; the reverse occurs in discharge. The substances used as electrodes and electrolytes vary from one type of battery to another. Lead-acid batteries are the most common, but more advanced types under development promise to achieve a higher specific energy (maximum stored energy per unit weight, measured in watt-hours per kilogram) and higher specific power (maximum power output per unit weight, measured in watts per kilogram).

First developed in 1859, lead-acid batteries are the workhorses of battery storage. They consist of two or more lead plates immersed in an electrolyte of dilute sulfuric acid. Normally the plates are covered with a layer of lead sulphate. As the battery charges, the surface of the positive plate (anode) is transformed into lead dioxide while the negative plate (cathode) is reduced to grey lead; discharging the battery reverses the process. Common lead-acid batteries can handle up to 500 charge/discharge cycles before losing the ability to hold a charge, while specialized ones have lifetimes of up to 2,000 cycles. Lead-acid battery efficiencies (energy output divided by input) can reach 80 percent. Those designed for electric vehicles have a specific energy of around 40 watt-hours per kilogram (Wh/kg) and a specific power of 70 to 100 watts per kilogram (W/kg), resulting in driving ranges of 60 to 100 miles at top speeds of 50 or 60 miles per hour, depending on the type of vehicle (McLarnon and Cairns 1989). Lead-acid batteries are the least expensive type of storage battery, but compared to some others they are bulky and heavy, and so in the long run are probably not appropriate for electric cars.

The largest lead-acid battery storage plant for utility applications is currently operated by Southern California Edison in Chino, California. The 8,256 batteries, which can provide four hours of storage at a peak output of 10 MW, are run five days a week during hours of heavy demand. Although originally built as a research project, the $13.5 million plant began commercial operation in January 1991 and will probably remain on-line until the batteries wear out. Maintenance costs have been somewhat higher than expected, running between $200,000 and $400,000 per year. A similar, but larger, lead-acid battery storage plant is planned for Puerto Rico (Von KleinSmid 1991).

By combining other metals and electrolytes, batteries can achieve a higher specific energy and specific power and longer cycle life than lead-acid batteries. The combinations under investigation include nickel-iron, zinc-bromine, zinc-chlorine and hydrogen-nickel oxide. The latter is preferred for use in space because of its long lifetime, but carries a price tag of roughly $20 per watt-hour of storage capacity. (By comparison, the Chino, California, battery facility costs $0.34 per watt-hour of storage capacity.) At present, most of the other advanced batteries are also more expensive than lead-acid batteries, but they are regarded as more promising for the long term.

A number of high-temperature batteries may be suited to mobile and stationary applications. One of these, the sodium-sulfur battery, uses a negative electrode of molten sodium, a positive electrode of molten sulfur-sodium polysulfide, and a solid sodium ion electrolyte that must be kept above 300°C to maintain sufficient electrical conductivity. Obviously, such batteries need to be thoroughly isolated from the environment to prevent (among other problems) explosive reactions with water. A sodium-sulfur battery of 50 kW peak output and 8 hours storage capacity developed for utility load-leveling applications has demonstrated an efficiency of 85 percent and a high specific energy, making this a leading candidate among advanced batteries for bulk electricity storage (McLarnon and Cairns 1989).

Pumped Hydroelectric Storage

Pumped hydroelectric storage is a large-scale storage technology that uses electricity to pump water from a lower reservoir to a higher one, then reproduces the energy by allowing the water to flow back through a turbine. Already widely used by electric utilities, as of 1991 more than 17,000 MW of pumped-storage capacity was in operation at 37 sites, with another 975 MW at two sites under construction. The Federal Energy Regulatory Commission has identified 53 other sites that could be developed with a potential capacity of 26,000 MW (FERC 1991). Sizes range from a few megawatts to more than 2,000 MW at the new Bath County plant in Virginia, while storage times range from about five hours to almost five days at the new Helms plant in California. System efficiencies typically run from 70 to 75 percent.

Like conventional hydropower, however, pumped storage faces serious environmental obstacles. Many of the best sites available for pumped storage, involving two adjacent valleys at different altitudes above sea level, lie in pristine wilderness areas. Consequently, opposition to pumped storage projects from environmental groups, sports fishermen, and others is often strong, and several proposed projects have been canceled. In addition to the usual concern over the prospect of inundating large land areas and destroying plant and animal habitats, objections are raised over the large daily water-level fluctuations that are seen in pumped-storage reservoirs and that interfere with recreational activities. Such problems, combined with a recent history of substantial cost overruns at projects, have made utilities reluctant to pursue pumped storage very aggressively (Boutacoff 1989a, Harza 1990).

At least some of these problems could be avoided by building the lower reservoir underground, in natural or excavated caverns, or even in abandoned gas wells. An underground pumped storage system is less intrusive but naturally costs more than an above ground system. To be worthwhile, it would probably have to have a capacity of at least 2,000 MW, and the distance between the upper and lower reservoirs would have to be greater than that usually required for conventional pumped storage (EPRI 1986).

The first underground pumped storage project is planned for Norton, Ohio, just southwest of Akron. This $1.7 billion project will use, as its lower reservoir, an old limestone mine capable of holding 338 million cubic meters of water. Water pumped at night to a reservoir created above ground will fall 680 meters (2,200 feet) through turbines to generate electricity during peak periods. Such a high head — the difference between the two water levels — allows for greater energy capture and efficiency, and thus a smaller upper reservoir (Carlson 1991).

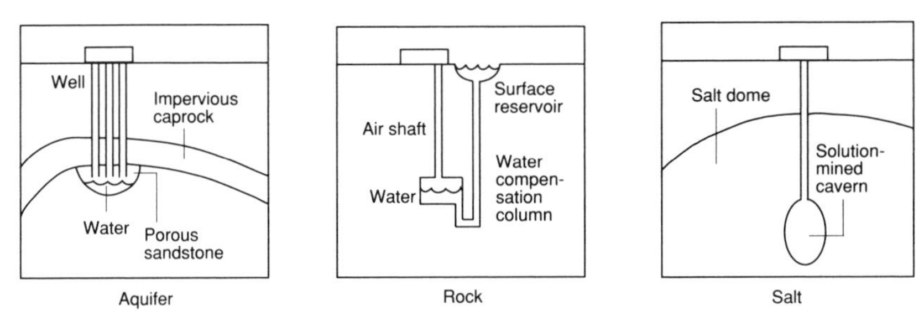

Figure 8.2
Compressed air storage plants are a promising option for large-scale, long-term electricity storage because of their modest cost, low environmental impacts, and wide geographic applicability. Of the three main types of storage media that can be used, salt caverns are the most thoroughly understood and aquifers the least expensive. Source: Electric Power Research Institute.

Compressed Air Energy Storage (CAES)

Just as energy can be stored by pumping water uphill, it can also be stored by compressing air into a confined space. To produce power, the compressed air is directed through a modified gas turbine, thus boosting the turbine's efficiency. This technology, only recently applied in the United States, offers a practical, low-cost alternative to pumped hydroelectric storage and carries few environmental impacts.

Four types of reservoir can be used to store compressed air: caverns mined from salt domes, and those excavated from bedrock, aquifers, and abandoned natural gas wells. Studies by the Electric Power Research Institute indicate that suitable areas for reservoirs of one type or another can be found in about three-quarters of the United States. Salt caverns of very large size are created by dissolving away salt in the middle of a dome, leaving an airtight chamber 100 to 1,000 meters below the surface. Suitable salt domes are found mainly in the Gulf of Mexico area. Hard rock is far more common and can be excavated in much the same manner as underground tunnels and pumphouses for pumped hydropower. To maintain pressure despite leaks, and to minimize the cavern volume and cost, the caverns would be connected to small water reservoirs at the surface (McKay, Steffens, and Curlin 1989). Both aquifers, which have been used for more than 50 years by natural gas companies for storing natural gas, and depleted natural gas wells can be converted relatively

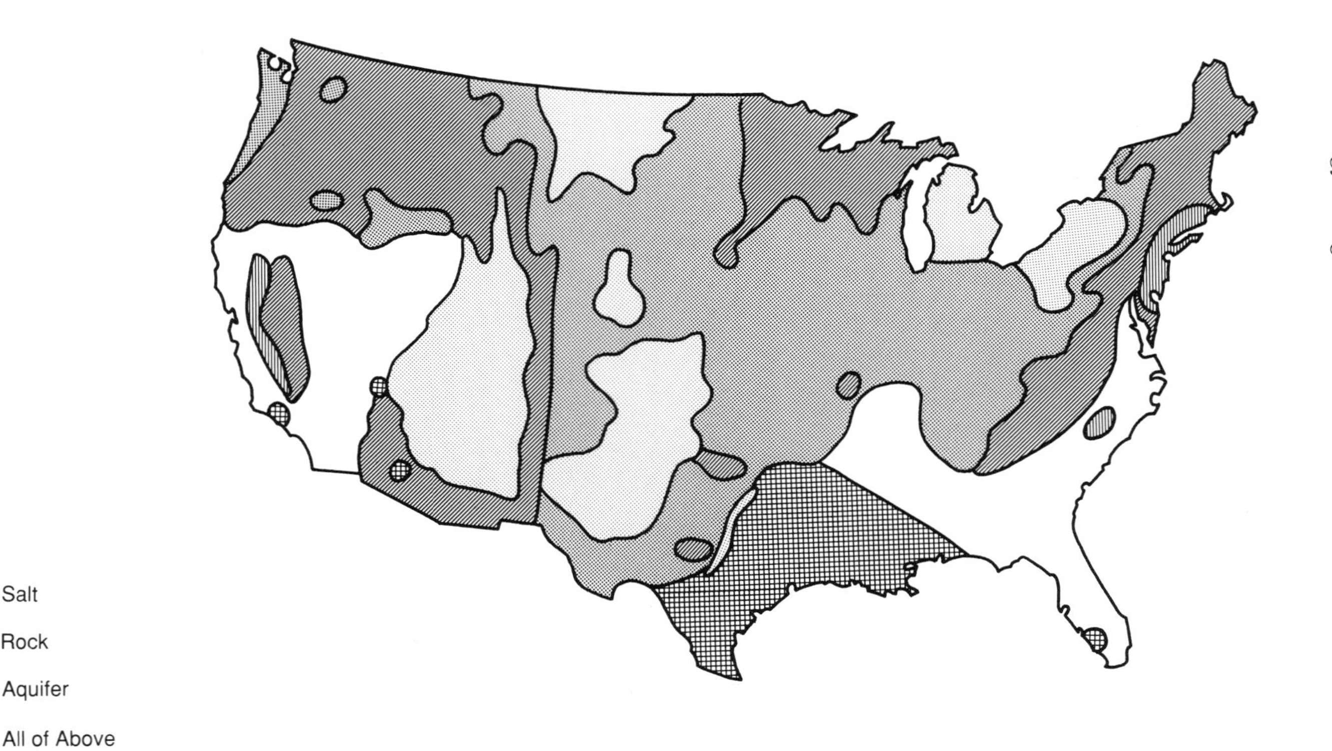

Figure 8.3
Map of the United States showing feasible areas for compressed air energy storage. Source: Electric Power Research Institute.

easily to compressed air storage and will consequently be the least expensive to develop.

Compressed air energy storage is a reality today. The first commercial CAES plant began operating in Huntorf, Germany, in 1978. This 290 MW plant, which stores compressed air in two salt caverns with a total volume of 300,000 cubic meters, is capable of generating at maximum power for up to four hours (Boutacoff 1989b). The first CAES plant in the United States was completed in 1991 for the Alabama Electric Cooperative, a rural utility. In this plant, off-peak electricity compresses air and stores it in a 540,000 cubic meter salt-dome reservoir whose bottom is 760 meters below the surface. When fully charged, air pressure inside the cavern is 75 bars, or 1,100 pounds per square inch (Burgess 1991). To generate power, the compressed air is first heated by gas burners, then passed through an expansion turbine. For every kilowatt-hour of electricity generated, the Alabama plant uses 0.82 kWh of electricity for compression and pumping and 3.9 MJ of gas. By comparison, conventional gas turbines require around 10 to 15 MJ to produce one kilowatt-hour of output.

Of course, because gas is used to drive the turbine, this technology is not pollution free. But the emissions are about three times lower than for comparable combustion turbines because of the higher efficiency of gas use. If salt domes are mined for the storage reservoir, the resulting dissolved brine may create problems of disposal, although innovative solutions are sometimes possible. (All of the brine generated during construction of Alabama Electric Cooperative's CAES plant, for example, was used by a nearby chemical company for the production of sodium, chlorine, hydrogen, and caustic soda.)

Superconducting Magnetic Energy Storage (SMES)

The most speculative electricity storage technology is superconducting magnetic energy storage (SMES). One of its attractions is that it provides the only known method of storing electricity directly, without first converting it into another form of energy, such as compressed air, and later converting it back. But while small test projects have shown SMES to be technically feasible, its present high cost and early stage of development make its future highly uncertain.

Electric current running through a loop of wire generates a magnetic field. If it were not for resistance in the wire, the current and magnetic field could remain there indefinitely; resistance causes the energy to

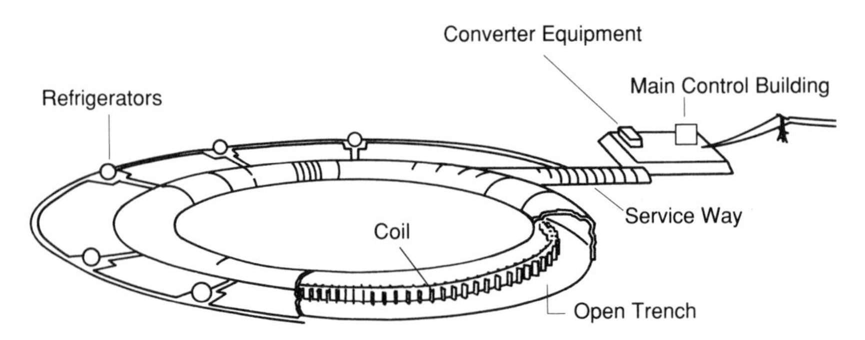

Figure 8.4
Artist's rendering of a utility-scale superconducting magnetic energy storage system. Source: Electric Power Research Institute.

dissipate as heat. For practical storage, it is necessary to make the coil out of wire with as close to zero resistance as possible: a superconductor. Presently virtually all commercial superconductors consist of metals such as titanium and niobium cooled to the very low temperature of liquid helium (-271°C). An intense effort is under way to develop practical high-temperature superconductors, which are materials that conduct electricity perfectly at or above the temperature of liquid nitrogen (-196°C). But until such materials are perfected, conventional superconducting materials will have to be used, implying the need for elaborate and costly cooling systems.

To achieve sufficient economies of scale, commercial SMES systems will probably need to have a peak power output of at least 1,000 MW and a storage capacity of at least 5 hours. At this scale, at least according to one study, SMES systems could be the lowest-cost option for utility-scale electricity storage. They would be too large for most single utilities, however, and so would have to be shared regionally (Giese and Rogers 1989).

In the early eighties, an experimental SMES system with a storage capacity of 8.3 kWh was built in Tacoma, Washington, to damp grid oscillations for the Bonneville Power Administration. After a successful year-long test, it was dismantled. The next SMES system could be a large one funded jointly by the Electric Power Research Institute and the Defense Nuclear Agency to supply high-power, ground-based lasers for military applications. Plans call for a system with 10 MW peak output and 5 hours storage capacity to be in operation by 1996 or 1997. The superconducting coil for this device would be approximately a kilometer in diameter and would be built into a rock trench for radial support. Liquid helium would bathe the coil, and an evacuated vessel would be built around the entire system to minimize heat gain. Storage efficiencies of 90 percent or more are expected, and the system is expected to be able to switch from charge to discharge mode in 20 milliseconds (EPRI 1990b, Birk 1992).

The intense magnetic field generated by a SMES coil is the most important environmental issue raised by this technology. Utility-scale plants are expected to generate magnetic fields of up to 5 Tesla, 100,000 times the strength of the earth's magnetic field (Blanchard 1989). While locally strong, these fields would not radiate and so would decrease rapidly with distance. Two kilometers from the center of the buried superconducting coil, the magnetic field would be only 0.001 Tesla, 20 times the earth's field (Cultu 1989). Higher fields are regularly generated in magnetic resonance imaging devices used in medicine and have not been shown to pose a health threat except to people with pacemakers or surgical clips (Blanchard 1989). Elevated magnetic fields could interfere with the navigation cues used by birds and insects to negotiate their home territories or for long-distance migration. These and other potential impacts must be given careful consideration before the first utility-scale system is built.

Hydrogen

For decades, visionaries have dreamed of a hydrogen economy, one powered by a clean, inexpensive, and inexhaustible fuel. The idea of producing hydrogen from solar- or wind-generated electricity and water, or from biomass, then using it to power vehicles, heat homes, and fuel industries, is attractive. The technology for generating hydrogen already exists, and millions of tons of hydrogen are used annually by U.S. industry to produce ammonia and methanol, refine petroleum products, hydrogenate edible oils, and lift rockets into space. Burning hydrogen releases only water and relatively small amounts of nitrogen oxides—no carbon dioxide, carbon monoxide, volatile organic compounds, or sulfur dioxide. And hydrogen is easily transported, either in liquid form or compressed and pumped through pipelines.

Despite these attractive qualities, hydrogen is not likely to come into wide use as a fuel source for at least two or three decades, and it will probably never entirely replace other alternative fuels. The major obstacles in the near term are the high cost of producing hydrogen, especially from renewable electricity sources, and the need to create a distribution network (similar to the natural gas pipeline network) to deliver it to customers. On a more fundamental level, hydrogen generated from electricity does not offer any clear advantages over electricity used directly, with the possible exception of electric vehicles, for which, as we will see, hydrogen fuel cells could prove an attractive alternative to batteries (Birk 1991).

Hydrogen atoms are almost always found combined with other atoms, and energy must be expended to break the chemical bonds and produce pure hydrogen gas (H_2). For this reason, hydrogen is properly thought of as an energy carrier or storage medium rather than a primary fuel such as sunlight or wind. Wind power, solar electricity, and hydropower are all well suited to generating hydrogen from the electrolysis of water. In electrolysis, electricity is passed directly through an electrolyte (water); oxygen is produced at the anode, hydrogen at the cathode. According to one study, hydrogen produced from photovoltaic electric-

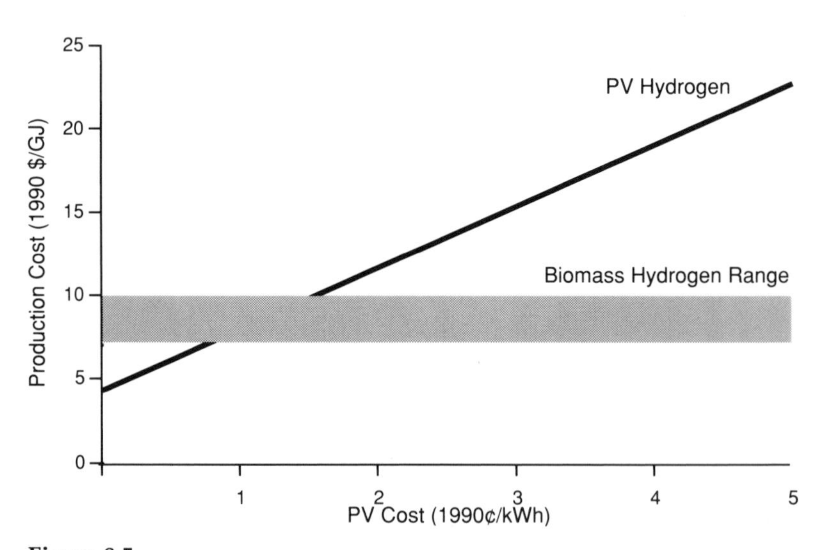

Figure 8.5
Hydrogen from photovoltaic (PV) electricity is likely to be cheaper than hydrogen derived from biomass only if the cost of PV electricity falls below 2¢/kWh. Sources: Ogden and Williams (1989); DeLuchi, Larson, and Williams (1991).

ity could be available at a competitive price within 10 years. By the year 2000, this study projects, mass-produced amorphous-silicon cells could cost as little as $0.16 per peak watt and achieve efficiencies as high as 18 percent, implying a delivered cost of hydrogen of around $8.40/GJ, roughly the current price of gasoline at the pump (Ogden and Williams 1989). Most other studies, however, do not forecast such a low cost of photovoltaic electricity in the near future (see chapter 3).

Studies by the Hawaii Natural Energy Institute and others indicate that an alternative approach, biomass gasification, could provide a much less costly source of renewable hydrogen in the near future. According to one analysis, a large plant consuming 8,000 tons of biomass daily could produce 44.5 billion standard cubic feet of hydrogen per year (about 116,000 tons) at a cost of $6.50/GJ (McKinley et al. 1990). The main drawbacks of using biomass as the feedstocks concern land use and other environmental impacts (discussed in chapter 5).

With only minor modifications, natural gas pipelines could transport compressed hydrogen from where it is produced to where it is consumed. The existing pipeline network radiates from the southern United States, where solar hydrogen production would likely be centered, to every part of the country. The hydrogen could then be used for all the same things for which natural gas is used, such as residential heating, industrial process heat, and electricity generation. If new pipelines have to be built to complement rather than replace the natural gas network — as seems likely in a transition from fossil fuels to hydrogen — the start-up cost of this distribution network could present a major obstacle.

Hydrogen is also being examined as a potential fuel for transportation, for which the main technical challenges are storage and conversion. In automobiles and trucks, hydrogen would either be compressed to pressures as high as 200 bars (3,000 pounds per square inch), liquefied, and kept refrigerated near -253°C, or stored as a metal hydride. Compressing hydrogen would be straightforward; a Canadian company, for example, is developing a bus powered by compressed hydrogen. For smaller cars, however, compressed hydrogen cylinders would be too heavy, so metal hydrides are a more promising approach. In this method, hydrogen is reversibly linked with metals such as nickel or iron to form a stable compound. When heated, the compound releases hydrogen, which is pumped to the engine cylinders and burned. The metal hydride storage batteries can be quickly recharged with compressed hydrogen and are relatively compact. For a car with an equivalent fuel efficiency of 40 kilometers per liter (100 miles per gallon), achieved by only a few very advanced prototypes today, and a driving range of 320 kilometers (200 miles), hydride storage batteries would weigh about 130 kilograms (280 pounds) and would occupy about as much space as a 12 gallon gasoline tank (Ogden and Williams 1989).

Hydrogen could also power vehicles through fuel cells, which convert chemical energy directly into electricity. Doing away with combustion would eliminate emissions entirely and would permit higher overall efficiencies to be attained. In some respects, fuel cells are similar to conventional batteries: A catalyst, like platinum, serves as the anode, which separates hydrogen into hydrogen ions and electrons. In a motor-vehicle fuel cell, the electrons would be routed into an electric motor, while the hydrogen ions would move to the cathode where they would combine with oxygen to form water.

A vehicle powered by hydrogen fuel cells was demonstrated in 1991, and others are likely to follow. According to one study, such vehicles would be pollution free, much more efficient than ones powered by internal combustion engines, and just as safe with respect to fire. In addition, the total cost of running the vehicle over its lifetime — including everything from fuel and maintenance to insurance and taxes — would be around $0.17 per kilometer ($0.27 per mile), about the same as for today's gasoline-powered cars (DeLuchi, Larson, and Williams 1991).

Because hydrogen is, naturally, highly inflammable, it poses some safety risks, although probably no greater than the risks associated with fossil fuels. Since hydrogen is colorless and odorless, additives must be used to make it more readily detected, just as is now done for natural gas. It also ignites more readily than gasoline or alcohol, and burns with a hot, almost invisible flame. On the other hand, hydrogen is a very light gas and so would disperse quickly in an accident, unlike petroleum products, which tend to remain in place.

Note

1. See chapter 4 for more detailed discussion.

9 Policies for a Renewable Future

As the foregoing chapters have shown, renewable energy technologies have made remarkable progress since the seventies and could begin to make a significant contribution to U.S. energy supply within the next decade or two. By the mid-nineties, wind power promises to be among the least expensive sources of electricity, fossil or renewable. Passive solar design strategies can already provide substantial energy savings in new buildings at little or no extra cost. Photovoltaic systems have come down 90 percent in cost and have been proven reliable in a wide range of applications from remote water-pumping stations to utility-scale power plants. Considerable progress has been made in improving the yields of energy crops and the conversion efficiencies for production of biofuels such as ethanol and syngas, which could become competitive with fossil fuels within five to ten years. Hydrothermal-geothermal technology is mature, and hot dry rock technology appears ready for full-scale demonstration, if not deployment. Although hydropower has limited room for expansion, ocean energy resources remain largely untapped, and the technologies required for this appear within reach.

And yet, the renewable energy industry is in the doldrums, with progress toward commercialization of nontraditional technologies slower now than at almost any time since the early seventies. To be sure, the picture is not entirely bleak. Wind energy development is moving ahead in several areas outside of California, sales of photovoltaic modules are increasing at an annual rate of about 20 percent, and even the troubled solar-collector industry has seen an increase in sales since the mideighties. Even so, it is difficult to escape the conclusion that these and other technologies are a long way from breaking into the mainstream economy. What is worse, they are given a very low priority in national and state energy policies. After more than a year of much-publicized hearings in which broad support for the use of renewable resources was indicated by the public, the Bush administration's National Energy Strategy provided few explicit policy measures (outside the realm of hydropower licensing) and no financial incentives to promote renewable energy development (DOE 1990e, NES 1991).

A Failure of Vision

Chapter 2 discussed some of the reasons that renewable energy technologies have trouble making headway in today's energy market. They range from a lack of information on available technologies and resources, to tax laws that tend to favor fuel-intensive technologies over capital-intensive ones, to the lack of consideration of environmental impacts.

These factors go a long way toward explaining the industry's present troubles but not the almost complete disinterest in renewable energy evident at the highest levels of government. This can only be ascribed to what might be called a failure of vision (Grubb 1990). Imagining a world in which renewable energy sources provide the bulk of energy supply means challenging the conventional wisdom and breaking out of well-worn patterns of thinking, a difficult hurdle for anyone to overcome. Perhaps just as important, for policymakers it also means challenging traditional energy industries with very deep pockets and a long history of close ties to government. There is consequently a strong tendency to remain committed to the status quo.

The failure of vision is by no means confined to the United States. Even in Japan, where government and industry are renowned for working hand-in-hand to promote their long-term economic interests, renewable energy technologies receive relatively little attention and funding, despite the very large renewable resources (such as geothermal energy) that exist there; in 1990, Japan actually spent less than the United States on renewable energy R&D as a percent of total energy-supply R&D. The European Community appears to be somewhat more forward-looking, and Germany in particular has seized the initiative to develop wind and photovoltaic technology for both domestic application and export. But no European country except Greece spent more than 25 percent of its energy-supply R&D funding in 1990 on renewable energy technologies (IEA 1991).

In trying to craft a strategy for expanding the use of renewable energy sources, one is faced with a chicken-and-egg problem. Policies to encourage renewable energy development can be defined, but it is difficult to see how they can be implemented without a fundamental change in the attitudes of policymakers. Yet the most direct way to change the attitudes of those in power is to demonstrate the success of nontraditional renewable energy technologies on a significant scale (and outside of California). But this will not happen anytime soon without vigorous new policies. And so on.

A Way Out?

The situation might well be hopeless were it not for one thing: the growing importance of environmental issues in the public consciousness and political debate. In the seventies, the principal motivating force behind the push to develop renewable energy was the oil "crisis." As soon as this crisis passed, the push disappeared. But concern about the environmental impacts of energy use increasingly pervades all levels of government and public discourse. It is seen in the virtual halt to nuclear power plant construction in the United States, the growing difficulty utilities face in building new coal-fired power plants, new federal laws to reduce emissions of acid-forming compounds and other pollutants, moves by several states to mandate electric cars, international agreements to ban chlorofluorocarbons, and high-profile international negotiations to restrict greenhouse-gas emissions. It is these manifestations of the public's growing environmental consciousness that, more likely than not, will force a reevaluation of federal and state energy policies.

This is not to say that renewable energy sources can provide the complete solution for the world's environmental ills. Biomass combustion produces air pollution, energy crops require large amounts of agricultural land, hydroelectric power plants can damage river ecosystems, and geothermal power plants often need large quantities of cooling water and face problems of waste disposal. Even the most benign renewable sources such as wind power can and have raised conflicts over land use. But if carefully managed, and in combination with high efficiency and, probably, greater use of natural gas, renewable energy sources could provide a way out of the emerging impasse between the growing demand for energy and the desire to protect our environment and quality of life.

The indispensable role of energy efficiency in making the transition from a fossil-fuel economy to one based on renewable energy sources is important to acknowledge. Renewable resources are large but not limitless, nor can they be expected to "catch up" to fossil fuels as long as energy demand continues to grow. To meet society's need for energy services — that is, power for transportation, manufacturing, space heating, communications, and other needs — from renewable sources with a minimum of environmental impacts will require the utmost in efficient energy use. Like passive solar building designs, which meld efficiency and solar technology into a single system, America's energy system must make use of both efficiency and renewable energy sources. To make the most of limited biofuel resources, for example, the fuel economy of automobiles must be increased; to reduce the total cost of solar water-heating systems, residential water use must be reduced; and to minimize the land required for wind, solar, and biomass generating facilities, the efficiency of electricity consumption in all sectors must be improved.

Assuming the nation can muster the political will to transform its energy system in this manner, many policies can be conceived of that would address one or another of the market barriers to renewable energy described in chapter 2. Choosing among them requires evaluating the policies against a number of criteria, chief among them effectiveness, cost, and ease of implementation. The policies outlined below are far from comprehensive, but they represent first steps to promoting greater use of renewable energy sources. They fall into six general categories: taxes, incentives, regulation, education, research, and environmental management.

Taxes

From the standpoint of economic efficiency, increasing taxes on fossil fuels is generally regarded as the best approach to reducing fossil-fuel consumption. It has the advantage that it would give energy consumers a clear price signal yet would allow them to change their consumption patterns in the most efficient and convenient manner for them. It would also provide a means to incorporate environmental and social costs of energy use directly in energy prices. The use of regulation to enforce changes, on the other hand, can sometimes lead to less-than-optimal consumer choices.

To affect overall energy consumption appreciably, however, the increase in fossil-fuel taxes would have to be quite large, probably comparable in magnitude to current fossil-fuel prices. But if the aim was merely to increase the incentive to purchase renewable energy sources, rather than to reduce overall consumption, then a smaller tax might be sufficient. For example, a nationwide tax increase of $0.50 per gallon of gasoline could accelerate by several years the introduction of biofuels such as ethanol, whose production cost from woody biomass is currently about $0.50 per gallon higher than the wholesale price of gasoline. Similarly, a tax of even 1¢/kWh on the price of electricity from coal-fired or gas-fired power plants would make today's wind technology far more widely competitive.

Choosing an appropriate level of tax is, clearly, a difficult task, both politically and analytically. One attractive option is to base the tax increases on some measure of the environmental and social costs of fossil fuels. For example, fuels could be taxed at a level corresponding to their carbon content, thus (in theory) optimizing consumer choices to minimize the risk of global warming. As we saw in chapter 2, however, estimates of environmental and social costs vary over a wide range.

Whether increasing taxes on fossil fuels would hinder economic growth is also a difficult question. Most macroeconomic models indicate it would place a heavy burden on the economy, but technology-oriented, bottom-up models tend to reach the opposite conclusion (see chapter 1). Recent unpublished analysis by the Environmental Protection Agency suggests that much depends on how the new tax revenue is spent (e.g., on reducing the federal deficit, on reducing other taxes, on tax credits for industry, or on new government programs) (Stevens 1992). The revenues could be extremely large: A tax of $25 per ton of carbon dioxide emitted would raise $140 billion annually, assuming continuation of current levels of fossil-fuel use (UCS 1991). One possible benefit of such taxes is that they would shift much of the present tax burden from productive activities (such as employment and capital gains) to polluting activities, thus stimulating economic growth while reducing stress on the environment (Dower and Repetto 1990).

Despite their attractiveness as an instrument of environmental policy, however, it will be very difficult to raise energy taxes to an appreciable degree in the present political atmosphere of the United States, even if the increase is linked directly to a decrease in other taxes. The situation is quite different in Europe, where the European Community, as of this writing, is preparing to implement a substantial carbon tax on fossil fuels. Until substantial new energy taxes are imposed in the United States, moving renewables forward in this country will require an emphasis on other policies, particularly financial incentives — such as those described below — targeted specifically at renewable sources.

Financial Incentives

To help renewable energy companies get off the ground, it is essential to create financial incentives designed to lower the initial costs and perceived financial risks associated with renewable energy technologies. In the late seventies and early eighties, the federal government and many states established incentives, such as tax credits, which in some cases were very successful in boosting sales of such products as wind turbines and solar collectors. In hindsight, however, these incentives were probably too generous and ill-focused and consequently contributed to some spectacular abuses and failures that have haunted the reputation of renewable energy ever since. Any new incentives must be carefully designed and monitored to avoid such problems. Four programs in particular could be effective:

- A performance-based tax credit of 2.5¢/kWh should be instituted for electricity produced from new renewable energy facilities.[1] Since the credit would be based on the amount of electricity produced rather than the amount of capital invested, it would not attract investors merely looking for a tax shelter (a problem with earlier, investment-based credits). The credit should be provided for a limited period of perhaps 7 to 10 years from the start of operation of a project to avoid the creation of a permanent subsidy. The credit should also be phased out for technologies as they become established and no longer require assistance.

- A similar tax credit of $2 per million Btu should be instituted for heat supplied to large industrial and commercial users from renewable sources. Both this credit and the previous one would supplant the existing 10 percent tax credit for solar and geothermal business investments, although companies could be allowed to claim the investment credit, in place of the new credit, for a limited period.

- Small businesses and homeowners not covered by the above tax credits should be allowed a 10 to 15 percent tax credit for renewable energy investments (e.g., solar water heaters and rooftop photovoltaic systems) meeting federal performance standards. Like the previous credits, it would be available for a limited period and would be phased out as technologies become more established.

- Government-financed loan banks and other similar programs should be established to help finance private investments in renewable energy products and facilities. To minimize risk, loans should be conditioned on an investment's meeting strict federal or state standards of performance and cost. An example of such a program is the proposal to allow the cost of solar water heaters certified by the Solar Rating and Certification Corporation to be included in mortgages approved by the Federal Housing Administration. Home buyers would thereby be able to spread the cost of the systems over 30 years while reaping immediate benefits in fuel savings. Other loan programs should be targeted at industry, utilities, and commercial users.

Regulation

As a rule, market-based strategies are preferable to regulation as a means to influence energy choices, but in some cases changes in regulation can be both effective and inexpensive. Here we focus on electric utilities, but initiatives will also be needed in the regulation of building construction, federal and state agencies, and transportation, among other areas (UCS 1991).

Electric utilities are regulated monopolies beholden to state commissions, which must approve electricity rates and power plant purchases through procedures that vary widely from state to state. The realm of electric-utility regulation consequently presents many opportunities for encouraging renewable energy use. At the moment, in most states, utilities have little or no incentive to investigate nontraditional supply sources, and even though they are required by law to purchase electricity from independent power producers if that is the least expensive option, unnecessary contract requirements and other details often present severe obstacles for companies hoping to supply that power from renewable sources (Moskovitz 1991).

The first step to encouraging renewable energy use is to require that the principles of "least-cost planning" be applied to all utility investments. This means that utilities must consider the full range of available resource options, including energy efficiency (also called demand-side management) and renewable energy, the goal being to provide consumers with electricity *services* — rather than electricity itself — at the lowest possible cost. As of 1991, only 14 states had fully adopted least-cost planning, but of the five states that had experienced the most success in developing renewable energy sources (other than hydropower), four had least-cost planning processes in place.

Least-cost planning is not enough, however. At least four additional steps should be adopted specifically to ensure that investments in renewable energy are given fair consideration:

- Electric utilities should be required to take environmental costs into account in evaluating new energy sources. Only a few states, including Nevada, Massachusetts, California, and New York, are seriously considering or have instituted regulations like this, although more states are likely to follow. One objection that is often raised is that because it is so difficult to calculate the environmental and social costs of energy, almost any estimates proposed by electric utilities or others will be open to challenge. Given the uncertainties, in fact, there will be a natural tendency for the external costs to be set too low, at least at first. Nevertheless, nearly any scheme would be preferable to the present one, in which these costs are implicitly assigned a value of zero.

- Existing restrictions mandated by PURPA (the Public Utilities Regulatory Policy Act) on the capacity of small power producers using renewable sources should be permanently lifted to permit plants to take advantage of economies of scale. For the time being, the capacity restriction of 80 MW has been waived by Congress, but this exemption will run out for those plants not certified as "qualifying facilities" before the end of 1994.

- Regulations should be instituted to ensure easier access to utility transmission by independent power producers of all types, ranging from industrial cogenerators to homes equipped with rooftop photovoltaic systems. Although access is already nominally a requirement of law, utilities often discourage it (for example, by charging high rates for backup power, setting unreasonable standards of power conditioning, and requiring expensive interconnection equipment).

- State utility commissions should require utilities to devote a significant fraction of their income to research and development on renewable alternatives and to consider such alternatives in planning future supply additions. U.S. utilities currently invest much less in research and development than do utilities in Europe and Japan, and little of the money that is spent goes to renewable sources. In part the fault for this short-sighted practice lies with state commissions, which tend to frown on utility investments that do not contribute tangibly to meeting consumer demand or reducing consumer costs. Thus, a change in attitude by both regulators and utility managers is required.

Education, Information, and Technical Assistance

Educating the public about the economic and environmental advantages of renewable energy sources and providing potential users with technical assistance to evaluate renewable alternatives are essential steps to overcoming the market's natural resistance to new technologies. Public education and technical assistance programs are relatively inexpensive but can have a large impact. Some programs are already funded by the Department of Energy and other federal and state agencies, but they could be greatly expanded, and efforts could be made to disseminate the products more widely.

For example, information brochures and booklets should be developed and distributed to homeowners and prospective home buyers describing options such as solar water heating and passive solar heating and cooling, providing guidelines to determine their cost effectiveness under various conditions, and suggesting ways to go about obtaining financing and locating qualified contractors. Design manuals and computer software have been developed for architects and building contractors to enable them to test prospective solar designs in residential and commercial buildings and evaluate their cost and performance, but these could be much more widely disseminated.

Analytical tools should be developed to enable electric utilities to evaluate renewable electricity sources more objectively. Since electric utilities first came into being, they have been accustomed to building and operating a few large power stations and adding transmission capability as needed to meet consumer demand. The introduction of numerous smaller, dispersed renewable facilities creates new problems and opportunities that most utilities are ill-equipped to evaluate. For example, because of the inadequacies of conventional planning methods, utilities are only now beginning to realize that grid-connected photovoltaic systems can be cost effective if they reduce peak loads on the transmission and distribution network, in addition to displacing conventional generation (see chapter 3).

Potential industrial and agricultural users of renewable energy should be provided with information and technical assistance suited to their particular needs. For example, farmers could be informed of cost effective ways to convert crop or animal wastes to energy through such technologies as anaerobic digestion. Renewable energy "success stories" could be widely publicized.

Lastly, a comprehensive system of performance standards and certification procedures should be instituted either by the federal government or by state agencies to give potential users of renewable energy confidence that the products they buy will indeed work as advertised. So far, what few standards and certifications exist — for solar water heaters, for example — have been established largely by industry with limited public funds and carry no government seal of approval.

Research and Development

Developing new technologies and ensuring their transfer to industry are key requirements for expanding the use of renewable energy sources. Current priorities in federal energy R&D funding are skewed, however, as most money goes to the development of advanced nuclear reactors, fusion, and "clean coal" technologies rather than energy-efficient and renewable energy technologies. For the next several years, federal renewable energy R&D funding should grow at a steady rate of perhaps 20-30 percent per year. (While a more rapid increase might seem desirable, it is doubtful that the additional funds could be efficiently absorbed and spent.) Within a few years a stable level of funding should be reached, perhaps at around $750 million to $1 billion per year (approximately the level of 1980).

Of course, funding more research and development will not necessarily lead to greater commercial success for renewable energy technologies if the funds are not well invested. While a certain number of mistakes and failures are a natural part of the R&D process, the Department of Energy renewable energy program has something of a history of backing losing technologies (such as multimegawatt wind turbines) and failing to support more successful technologies (such as solar parabolic trough systems).

Perhaps the best way to avoid such problems in the future is to work more closely with industry to establish realistic research goals and priorities, something the Department of Energy is already beginning to do in a small way. In addition, the Department of Energy should seek to strike a balance between basic research on long-term technologies and commercialization of near-term technologies. To this end, a substantial portion of R&D funding should be set aside for cost-shared government-industry demonstration projects, which were cut to the bone in the past decade. One model of such a collaboration is the Photovoltaics for Utility-Scale Applications (PVUSA) project, a five-year program involving side-by-side testing of photovoltaic modules in a working utility environment. Another is the effort of scientists at the National Renewable Energy Laboratory and various companies to develop biomass-derived replacements for phenol, a chemical used in adhesives, resins, and other products. Similar efforts are needed to evaluate the commercial feasibility of, among other technologies, hot dry rock, the biomass integrated gasifier/gas turbine, large-scale biomass cultivation, biofuel production, and other promising concepts.

One area of research that is perennially underfunded is resource surveys and assessments. Solar radiation data have been recorded consistently at only a relatively small number of sites around the country, data on wind speeds are similarly sparse and often not taken at the appropriate height for modern intermediate-size wind turbines, and hydrothermal reservoirs and thermal gradients for hot dry rock have not been systematically investigated. This lack of data can present a virtually insurmountable obstacle for small companies that do not have the money to conduct detailed resource assessments on their own. It also often leaves state and local energy planners without a clear picture of the possibilities for renewable energy development in their region.

Because renewable energy resources vary so widely by region, state and local efforts could play a useful role in advancing their development. In agricultural states, for example, funds could be invested in projects to demonstrate integrated energy crop and biofuels or electricity production. Until markets for biofuels develop, the products could be sold to state and federal agencies to power fleet vehicles. In states with especially good wind resources, pilot wind projects could be developed to give utilities experience with wind power. States in the south and southwest have an obvious interest in the development of solar energy for buildings, industry, and utilities. The potential economic and environmental benefits of such projects for the states concerned should be made clear. Not only would the environment be improved, but the groundwork would be laid for the establishment of new local industries and job markets that could help sustain state economies into the next century.

Environmental Regulation and Monitoring

As much as it is desirable to promote the development of renewable energy sources, it is also necessary to take steps to ensure that their environmental impacts are minimized. Although the impacts are generally less severe than those of fossil fuels, they can be significant, especially for energy crops, biomass and municipal solid waste incineration, and hydropower. Here are several issues that should be considered:

• The desire to consolidate and "streamline" the licensing of hydroelectric facilities (a proposal of the National Energy Strategy) must be balanced against the possible damage to river ecosystems and loss of scenic and recreational value that could result from insufficient oversight and involvement by the public and appropriate government agencies. Although some consolidation of state and federal hydropower licensing procedures seems sensible, it should not come at the cost of ignoring public concerns or weakening environmental standards.

• The growing of energy crops such as grasses, woody shrubs, and short-rotation trees should be carefully monitored and regulated to minimize impacts to land, water, and wildlife. Environmental guidelines should be developed that take into account such factors as the quality of land (e.g., susceptibility to erosion), types of energy crops to be grown, the amount of artificial fertilizer, pesticides, and herbicides to be used, and tilling and harvesting practices. In general, perennial grasses and short-rotation trees should be preferred over annual crops because their cultivation causes much less erosion.

• Burning municipal solid wastes to generate electricity can produce air pollution (including sulfur dioxide, nitrogen oxides, hydrochloric acid, and a variety of toxic metals) in significantly higher quantities than other forms of electricity generation. In addition, it produces ash that is often high in soluble toxics and must be disposed of safely. Because of these problems, states and cities should take all practical steps to encourage source reduction, recycling, and reuse of waste materials before considering the option of incineration. Where incineration cannot be avoided, state and federal regulations should ensure that the best technology is used to control pollutant emissions and toxic ash disposal.

Other issues that should be addressed include the recycling or disposal of toxic materials used in batteries, photovoltaic modules, and other systems; establishing guidelines for the siting and operation of wind turbines (to protect endangered bird species, for example); and setting pollution standards for automobiles burning alcohol fuels.

Conclusions

The policy agenda outlined here represents only a first step toward a renewable future for the United States. Some of these policies are obviously necessary; others must be tested for effectiveness and cost. Even assuming this full program is implemented, additional steps may still be needed. Nevertheless, something like this program will have to be enacted if the United States is to reap the full benefits of the remarkable technology that has been developed in the past decade.

In any case, we do not have much choice. If nothing is done, then within a decade or two a massive shift away from fossil fuels could well be forced upon us, either by some crisis in world energy supply stemming from diminishing reserves of oil and natural gas, or by the belated realization that greenhouse warming is a reality that will mean disaster for much of humanity. Then, Americans may well wonder why government leaders ignored the signs and failed to act in a responsible manner.

A far better course is open to us, and a far better world is waiting, if we begin now to develop the available renewable resources.

Note

1. A similar proposal was originally included in the draft National Energy Strategy report submitted by the Department of Energy to the White House but was later dropped. The size of that credit (2¢/kWh) is reported to have been chosen to reflect unfair treatment of capital-intensive renewable technologies under existing tax codes. A 2.5¢/kWh production incentive later appeared in proposed legislation in the House and Senate but as of this writing had not been approved.

Appendix A: Units and Conversion Factors

By and large, this book uses metric units. The base unit of energy consumption is consequently the joule (J); one joule approximately equals the energy required to raise a kilogram weight 10 centimeters. National energy consumption is expressed in exajoules (EJ), where one exajoule equals 10^{18} joules; the United States consumes about 89 exajoules annually. The cost of energy (other than electricity) is expressed in dollars per gigajoule (GJ), where a gigajoule is 10^{9} joules; a gasoline price of $1 per gallon is equivalent to an energy cost of about $8 per gigajoule.

The base unit of electrical power is the watt (W), which equals a joule per second. The capacity of electric power plants is expressed in kilowatts (kW) or megawatts (MW), the quantity of electricity produced in kilowatt-hours (kWh), and the price in cents per kilowatt-hour (¢/kWh). One kilowatt-hour equals 3.6 megajoules (MJ), but because of conversion losses about 10.9 megajoules of so-called primary energy are actually needed to generate a kilowatt-hour of electricity in a conventional steam-electric power plant. The latter number is used to calculate the quantity of fossil fuels displaced by renewable and nuclear sources of electricity. For example, wind turbines generated 2.7 billion kilowatt-hours of electricity in 1991, and so displaced the fossil-fuel (or primary-energy) equivalent of 0.03 exajoules.

For those more familiar with English units, a British thermal unit, or Btu, is equal to 1055 joules; thus, an exajoule is almost exactly equal to a quadrillion Btu, or quad, and a gigajoule is almost exactly equal to a million Btu. The following tables provide some approximate equivalents of metric energy units and conversion factors.

Table A.1
Approximate energy equivalents.

1,000 joules (J)	1 match tip
1 gigajoule (GJ)	90 pounds of coal 120 pounds of oven-dried hardwood 8 gallons of gasoline 1,000 standard cubic feet of natural gas
1 exajoule (EJ)	45 million short tons of coal 60 million short tons of oven-dried hardwood 170 million barrels of crude oil 4 days of U.S. energy use 26 days of U.S. gasoline use

Table A.2
Conversion factors.

Length	
1 meter (m)	3.28 feet (ft)
1 kilometer (km)	0.621 miles
Area	
1 m^2	10.75 ft^2
1 km^2	0.386 $mile^2$
	100 hectares
1 hectare (ha)	2.47 acres
Weight	
1 kilogram (kg)	2.2 pounds
1 metric ton	1,000 kg
	1.1 short tons
Time	
1 hour (h)	3600 s
1 year (yr)	8760 h
Energy	
1,000 J	0.948 Btu
1 GJ	0.948 million Btu
1 EJ	0.948 quadrillion Btu (quad)
Energy density	
1 MJ/m^2	88 Btu/ft^2
Power	
1 kilowatt (kW)	1,000 J/s
Flux (power/unit area)	
1 $watt/m^2$	0.317 $Btu/ft^2/hr$
Temperature	
°C	(°F-32) x 5/9

Appendix B: U.S. Renewable Energy Funding

Department of Energy Renewable Energy R&D Funding ($ Millions).

	FY74	FY75	FY76	FY77	FY78	FY79	FY80	FY81	FY82
Solar									
Buildings	6.9	16.2	55.2	100.6	127.6	96.0	85.2	73.4	24.1
Photovoltaics	3.6	5.0	28.9	50.9	76.6	118.8	148.6	151.6	61.6
Solar Thermal	4.4	13.2	33.3	67.1	104.1	109.3	117.3	134.6	42.4
Biofuels	2.4	1.5	6.1	9.5	20.8	42.4	51.2	49.7	30.1
Wind	1.8	7.9	19.3	24.6	35.3	59.6	60.1	77.5	16.6
Ocean Energy	1.2	2.8	8.6	14.5	31.2	41.1	43.0	34.6	20.8
International								10.8	4.0
Tech Transfer								1.4	6.7
NREL*						3.0	6.9	5.0	
Resource Assessment									
Program Direction				1.5	2.5	3.4	2.9	6.8	4.0
Program Support									3.5
Other	0.6	1.4	7.3	13.5	10.7	10.7	33.2	3.5	7.5
SUBTOTAL	20.8	48.0	158.7	282.2	408.8	484.3	548.4	548.9	221.3
Geothermal		27.1	40.7	54.7	106.2	155.2	149.2	156.0	56.4
Small Hydro					10.7	26.0	20.9	3.2	-2.9
TOTAL	20.8	75.1	199.4	336.9	525.7	667.5	718.5	708.1	274.8

* National Renewable Energy Laboratory, formerly the Solar Energy Research Institute (SERI).

Source: Fred J. Sissine, *Renewable Energy: A New National Commitment?* Congressional Research Service Issue Brief, January 6, 1992.

	FY83	FY84	FY85	FY86	FY87	FY88	FY89	FY90	FY91	FY92	TOTAL
	10.8	16.6	9.2	7.9	6.0	5.4	5.3	4.1	2.0	2.0	654.5
	57.9	50.2	54.5	37.8	40.0	34.6	35.1	34.7	46.3	60.4	1097.0
	48.5	38.7	33.9	25.5	22.6	17.0	14.8	15.0	19.3	29.1	890.1
	19.5	28.3	29.9	27.0	23.8	17.1	13.2	16.2	33.1	39.3	461.1
	31.4	26.4	28.4	27.3	16.5	9.7	8.7	9.1	11.1	21.4	492.7
	10.5	5.7	4.0	4.9	4.4	3.3	4.1	4.1	2.7	2.0	243.5
	10.0	0.5	0.4	1.0	0.8	0.8	1.0	1.0	1.5	2.0	33.8
	3.0	3.3	4.9	3.2	2.5	2.6	2.4	1.8	2.2	1.0	35.0
				2.0	0.5	0.6	0.7	0.6	5.1	11.5	35.9
					0.6	0.7	0.8	0.8	1.0	1.2	5.1
	5.9	6.0	3.5	4.3	4.1	4.2	4.4	4.1	4.3	4.7	66.6
	0.4	0.8	1.0	1.3	0.7	0.9	1.0	0.9	0.9	0.9	12.3
	-0.9		0.1	1.0							88.6
	197.0	176.5	169.8	143.2	122.5	96.9	91.5	92.4	129.5	175.5	4116.2
	56.6	30.3	31.3	26.5	20.6	20.9	19.3	18.1	27.3	27.2	1023.5
	1.0	0.8	0.1	0.5	0.5				1.0	1.0	64.8
	254.6	207.6	201.1	170.2	143.6	117.8	110.8	110.5	157.8	203.7	5204.5

Suggested Readings

The field of energy policy, and renewable energy in particular, has a rich literature that cannot be summarized in a few paragraphs. Here I suggest several sources of information, however, that should give the interested reader a good start.

The heyday of writing on energy issues was the late seventies, when solving the "crisis" of America's dependence on foreign oil was at the top of the national agenda. Among the many books emerging from that period arguing for major changes in energy policy, three stand out: *Energy Future*, edited by Robert Stobaugh and Daniel Yergin (New York: Random House, 1979), *Energy Strategies: Toward a Solar Future*, edited by Henry W. Kendall and Steven J. Nadis (Cambridge, Mass.: Ballinger, 1980), and *Soft Energy Paths*, by Amory B. Lovins (New York: Penguin Books, 1977). Among the reviews of renewable energy technologies, one of the best was *Solar Energy in America*, by W.D. Metz and A.L. Hammond (Washington, D.C.: American Association for the Advancement of Science, 1978).

Readers interested in learning more about air pollution, global warming, and other energy-related environmental problems have many books and reports to choose from. Probably the best way to survey this field is to review two reports published annually, the *World Resources* series produced by the World Resources Institute (1735 New York Avenue N.W., Suite 400, Washington, DC 20006), and the *State of the World* series produced by the Worldwatch Institute (1776 Massachusetts Avenue N.W., Washington, DC 20036). The latter organization also publishes the *Worldwatch Papers* series and the magazine *Worldwatch*, which are worth subscribing to. For more technical articles on energy and environmental policy there is no better resource than the *Annual Review of Energy and the Environment*, published by Annual Reviews (4139 El Camino Way, Palo Alto, CA 94303-0897).

Unfortunately, rather few in-depth studies of the status and prospects of renewable energy technologies have been published in the last several years. The comprehensive *New Electric Power Technologies*, published in 1985 by the U.S. Congress Office of Technology Assessment, contains sections on renewable energy sources, and an updated version should be available soon after this book goes to press. An extensive review of renewable energy technologies is provided in the proceedings of the conference *Energy and the Environment in the 21st Century*, edited by Jefferson W. Tester, David O. Wood, and Nancy A. Ferrari (Cambridge, Mass.: MIT Press, 1990). Another valuable reference is *The Potential of Renewable Energy: An Interlaboratory White Paper*, available from the National Renewable Energy Laboratory (1617 Cole Boulevard, Golden, CO 80401).

For a more concise (though uncritical) snapshot of renewable energy technologies, the reader should investigate Nancy Rader's *Power Surge*, published in 1989 by Public Citizen (215 Pennsylvania Avenue S.E., Washington, DC 20003), and a companion report by the same author, *The Power of the States*, published in 1990.

The best way for readers to keep abreast of developments in the renewable energy field is to contact some of the many organizations that regularly publish journals, reports, and other information. One source is the Department of Energy's annual renewable energy program reviews, which are available from the National Renewable Energy Laboratory. The California Energy Commission (1516 Ninth Street, Sacramento, CA 95814) publishes the useful *Energy Technology Status Report*, as well as other reports. Numerous helpful fact sheets and reports are available from the Electric Power Research Institute, Storage and Renewables Department (3412 Hillview Avenue, Palo Alto, CA 94303).

The American Solar Energy Society (2400 Central Avenue, B-1, Boulder, CO 80301) publishes reports and conference proceedings as well as the journal *Solar Today*. Renew America, a nonprofit advocacy group, produces colorful and informative booklets. The various renewable industry trade groups based in Washington, D.C. (the Solar Energy Industries Association, the American Wind Energy Association, the National Wood Energy Association, the National Hydropower Association, and others) are storehouses of valuable (if one-sided) information and also publish newsletters and journals. The Geothermal Resources Council (P.O. Box 1350, Davis, CA 95617-1350), which publishes the *Bulletin* and *Transactions*, serves much the same function for the geothermal industry. The Union of Concerned Scientists (26 Church Street,

Cambridge, MA 02238) produces a range of educational materials, including teaching and organizing manuals, brochures, and longer briefing papers and reports.

Anyone interested in renewable energy sources for home or small-business use should order the *Alternative Energy Sourcebook*, a catalog of products published by Real Goods Trading Corporation (966 Mazzoni Street, Ukiah, CA 95482).

References

Aitken, Donald W. 1992. Union of Concerned Scientists. Personal communication.

Alson, Jeffrey. 1990. "The Methanol Debate: Clearing the Air." In *Methanol as an Alternative Fuel Choice: An Assessment*. Edited by Wilfrid L. Kohl. Washington, D.C.: Johns Hopkins Foreign Policy Institute.

American Paper Institute (API). 1991. "U.S. Pulp and Paper Industry's Energy Use." New York: American Paper Institute.

Anderson, David N., and John W. Lund. 1987. "Geothermal Resources." *Encyclopedia of Physical Science and Technology*, vol. 6. New York: Academic Press.

Anderson, J. Hilbert. n.d. "Ocean Thermal Power: The Coming Revolution." York, Pa.: Sea Solar Power.

Armstead, H. Christopher. 1983. *Geothermal Energy*. London: E. & F.N. Spon.

Armstead, H. Christopher, and Jefferson W. Tester. 1987. *Heat Mining*. London: E. & F.N. Spon.

Awerbuch, Shimon. 1991a. "Measuring the Cost and Benefits of New Technology: The Case of Photovoltaics." Lowell, Mass.: University of Lowell. Draft.

———. 1991b. "The Role of Risk and Discount Rates in Utility Integrated Resource Planning." Lowell, Mass: University of Lowell.

Bankston, Charles A., and Dwayne S. Breger. n.d. "Description and Status of Central Solar Heating Plants with Seasonal Storage." Washington, D.C.: CBY Associates.

Beranyi, E. 1991. Government Advisory Associates, Inc. Personal communication.

Bergey, Michael. 1990. "Comments on the Maturation and Future Prospects of Small Wind Turbine Technology." Presented at the American Solar Energy Society Solar 90 Conference. Norman, Okla.: Bergey Windpower.

———. 1991. President, Bergey Windpower. Personal communication.

Bergoust, D. 1990. "Thermal Storage — The Swimming Pool as an Option." *Energy Engineering* 87, no. 6 (June): 46–56.

Bernow, Stephen S., Marc Breslow, Monica Becker, Kevin Gurney, Donald Marron, Daljit Singh, and John Stutz. 1990. *Incorporating Environmental and Economic Goals into Nevada's Energy Planning Process*. Boston: Tellus Institute.

Biologue. 1991. "Garbage Produces High-Quality Methane Gas in a Solar Energy Research Institute/Walt Disney World Project." *Biologue* 8, no. 3 (September): 13.

Birk, James R. 1991. Testimony to the U.S. Senate, Subcommittee on Energy Research and Development, Committee on Energy and Natural Resources. June 25.

———. 1992. Storage and Renewables Department, Electric Power Research Institute. Personal communication.

Blanchard, Janie Page. 1989. "Environmental Issues Associated with Superconducting Magnetic Energy Storage (SMES) Plants." *Proceedings of the 24th Intersociety Energy Conversion Engineering Conference* 4: 1777–1782. CH2781-3/89. New York: Institute of Electrical and Electronics Engineers.

Boes, Eldon. 1992. National Renewable Energy Laboratory. Personal communication.

Boley, Gary L. 1990. "Resource Recovery: Turning Waste into Watts." *Mechanical Engineering* 112, no. 12 (December): 37–41.

Bolin, Bert, Bo R. Doos, Jill Jager, and Richard A. Warrick. 1986. *The Greenhouse Effect, Climatic Change, and Ecosystems*. New York: J. Wiley and Sons.

Bony, Paul. 1992. Sierra Pacific Power Company. Personal communication.

Boutacoff, David. 1989a. "Emerging Strategies for Energy Storage." *EPRI Journal* 14, no. 5 (July/August): 4-13.

———. 1989b. "Pioneering CAES for Energy Storage." *EPRI Journal* 14, no. 1 (January/February): 30–39.

Breger, D.S., J.E. Sunderland, and H. Elhasnaoui. 1991. "Preliminary Design Development of a Central Solar Heating Plant with Seasonal Storage at the University of Massachusetts, Amherst." In *1991 Solar World Congress Proceedings*, vol. 1, part 1. Edited by M.E. Arden, Susan M.A. Burley, and Martha Coleman. New York: Pergamon Press.

Broadman, H., and W. Hogan. 1988. "Is an Oil Tariff Justified? The Numbers Say 'Yes.'" *Energy Journal* (July).

Brown, Donald W. 1990. "Using HDR Technology to Recharge The Geysers." *Proceedings: The National Energy Strategy — The Role of Geothermal Technology Development*. CONF-9004131. Springfield, Va: National Technical Information Service.

Bull, Stanley R. 1991. "The U.S. Department of Energy Biofuels Research Program." In *Energy from Biomass and Wastes XIV*. Edited by Donald L. Klass. Chicago: Institute of Gas Technology.

Bureau of the Census. 1990. *Statistical Abstract of the United States: 1990* (110th Edition). Washington, D.C.: U.S. Department of Commerce.

Bureau of Reclamation. 1991. *Hydropower 2002: Reclamation's Energy Initiative*. Washington, D.C.: U.S. Department of the Interior.

Burgess, Phillip. 1991. Alabama Electric Cooperative. Personal communication.

California Energy Commission (CEC). 1987. *Relative Cost of Electricity Production*. P300-86-006. Sacramento, Calif.

———. 1988. *Energy Technology Status Report*. Sacramento, Calif.

Cannon, James S. 1990. *The Health Costs of Air Pollution: A Survey of Studies Published 1984–1989*. Washington, D.C.: American Lung Association.

Carlson, David E. 1989. "Low-Cost Power From Thin Film Photovoltaics." In *Electricity: Efficient End-Use and New Generation Technologies and Their Planning Implications*. Edited by T.B. Johansson, B. Bodlund, and R.H. Williams. Lund, Sweden: Lund University Press.

Carlson, Robert. 1991. Consolidated Hydropower. Personal communication.

Caruso, Lisa. 1991. "Lights Help Save the Shad." *York* (Pa.) *Dispatch*, October 16.

Chandler, William U., Howard S. Geller, and Marc R. Ledbetter. 1988. *Energy Efficiency: A New Agenda*. Washington, D.C.: American Council for an Energy-Efficient Economy.

Charlier, R.H. 1982. *Tidal Energy*. New York: Van Nostrand Reinhold.

Chem Systems. 1989. "Assessment of Cost of Production of Methanol from Biomass." Golden, Colo.: National Renewable Energy Laboratory.

Cheremisinoff, Paul N., and Thomas C. Regino. 1978. *Principles and Applications of Solar Energy*. Ann Arbor, Mich.: Ann Arbor Science Publishers.

Chernich, P., and E. Caverhill. 1991. "Methods of Valuing Environmental Externalities." *Electricity Journal* 4, no. 2 (March): 46–53.

Chu, T.Y., J.C. Dunn, J.T. Finger, J.B. Rundle, and H.R. Westrich. 1990. "The Magma Energy Program." Geothermal Resources Council *Bulletin* 19, no. 2 (February): 42–52.

Chum, Helena L. 1990. "Inexpensive Phenol Replacements from Biomass: An On-Going Technology Transfer Effort." Golden, Colo.: National Renewable Energy Laboratory.

Cleveland, Cutler J., and Robert Herendeen. 1989. "Solar Parabolic Collectors: Successive Generations Are Better Net Energy and Exergy Producers." *Energy Systems and Policy* 13: 63–77.

Cohen, R. 1982. "Energy from the Ocean." *Philosophical Transactions of the Royal Society of London* 307A (October): 405–437.

———. 1992. Personal communication.

Conrad, David. 1991. "Regarding the Hydropower Regulating Aspects of S. 341." Testimony before the U.S. Senate, Committee on Energy and Natural Resources. Washington, D.C.: National Wildlife Federation. February 26.

Cook, James H., Jan Beyea, and Kathleen H. Keeler. 1991. "Potential Impacts of Biomass Production in the United States on Biological Diversity." *Annual Review of Energy and the Environment* 16: 401–431.

Cultu, M. 1989. "Superconducting Magnetic Energy Storage." In *Energy Storage Systems*. Edited by B. Kilkis and S. Kakac. Norwell, Mass.: Kluwer Academic Publishers.

DeLuchi, Mark A., Eric D. Larson, and Robert H. Williams. 1991. "Hydrogen and Methanol: Production from Biomass and Use in Fuel Cell and Internal Combustion Engines." Princeton, N.J.: Center for Energy and Environmental Studies, Princeton University.

Department of Energy (DOE). 1988a. *Programs in Renewable Energy: Fiscal Year 1989*. DOE/CH10093-38. Golden, Colo.: National Renewable Energy Laboratory.

———. 1988b. *Five-Year Research Plan 1988–1992: Biofuels and Municipal Waste Technology Program*. DOE/CH10093-25. Golden, Colo.: National Renewable Energy Laboratory.

———. 1989. *Geothermal Program Summary*. DOE/CH10093-49. Golden, Colo.: National Renewable Energy Laboratory.

———. 1990a. *Biofuels Program Summary*. DOE/CH10093-68. Golden, Colo.: National Renewable Energy Laboratory.

———. 1990b. *Solar Buildings Program Summary*. DOE/CH10093-62. Golden, Colo.: National Renewable Energy Laboratory.

———. 1990c. *Solar Thermal Program Summary*. DOE/CH10093-60. Golden, Colo.: National Renewable Energy Laboratory.

———. 1990d. *Ocean Energy Program Summary*. DOE/CH/10093-66. Golden, Colo.: National Renewable Energy Laboratory.

———. 1990e. *Interim Report, National Energy Strategy: A Compilation of Public Comments*. DOE/S-0066P. Washington, D.C.: Government Printing Office.

———. 1991a. *Electricity from Biomass: A Development Strategy*. Golden, Colo.: National Renewable Energy Laboratory. Draft.

———. 1991b. *Photovoltaics Program Plan: FY 1991–FY 1995*. DOE/CH10093-92. Golden, Colo.: National Renewable Energy Laboratory.

DiPippo, Ronald. 1990. "Geothermal Energy: Electricity Production and Environmental Impact, a Worldwide Perspective." In *Energy and the Environment in the 21st Century*. Edited by Jefferson W. Tester, David O. Wood, and Nancy A. Ferrari. Cambridge, Mass.: MIT Press.

Dodge, Darrell. 1992. National Renewable Energy Laboratory. Personal communication.

Dostrovsky, Israel. 1991. "Chemical Fuels from the Sun." *Scientific American* 265, no. 6 (December): 102–107.

Dower, R., and R. Repetto. 1990. "Use of the Federal Tax System to Improve the Environment." Testimony before the U.S. House of Representatives Committee on Ways and Means. March 6.

Dowling, James. 1991. "Hydroelectricity." In *The Energy Sourcebook*. Edited by Ruth Howes and Anthony Fainberg. New York: American Institute of Physics.

Dubin, Jeffrey A., and Geoffrey S. Rothwell. 1990. "Subsidy to Nuclear Power Through Price-Anderson Liability Limit."*Contemporary Policy Issues* 8, no. 3 (July): 73–79.

Duchane, David. 1991. "International Programs in Hot Dry Rock Technology Development." Geothermal Resources Council *Bulletin* 20, no. 5 (May): 135–142.

———. 1992. Manager, Hot Dry Rock Program, Los Alamos National Laboratory. Personal communication.

Dugger, G.L., E.J. Francis, and W.H. Avery. 1978. "Technical and Economic Feasibility of Ocean Thermal Energy Conversion." *Solar Energy* 20: 259–274.

Eichelberger, John C., and James C. Dunn. 1990. "Magma Energy: What Is the Potential?" Geothermal Resources Council *Bulletin* 19, no. 2 (February): 53–56.

Electric Power Research Institute (EPRI). 1986. "Pumped Hydro: Backbone of Utility Storage." *EPRI Journal* 11, no. 1 (January/February): 25–31.

———. 1989a. "Molten Salt Solar-Electric Experiment: Testing Operation and Evaluation." EPRI GS-6577. Palo Alto, Calif.

———. 1989b. "Pilot Test of the Crushed-Rock Storage Furnace: Summary of Findings." EPRI EM-6144. Palo Alto, Calif.

———. 1989c. *Technical Assessment Guide*, vol. 1 (Electricity Supply). EPRI P-6587-L. Palo Alto, Calif.

———. 1990a. "Excellent Forecast for Wind." *EPRI Journal* 15, no. 4 (June): 15–25.

———. 1990b. "Superconducting Magnetic Energy Storage Pilot Plant." EPRI TB.GS.90.2.90. Palo Alto, Calif.

———. 1991. "On-Site Utility Applications for Photovoltaics." *EPRI Journal* 16, no. 2 (March): 27–37.

Eliot Allen & Associates. 1980. *Preliminary Inventory of Western U.S. Cities with Proximate Hydrothermal Potential*, vol. 1. Klamath Falls, Ore.: Geo-Heat Center, Oregon Institute of Technology.

Elliott, D.L., L.L. Wendell, and G.L. Gower. 1991. *An Assessment of the Available Windy Land Area and Wind Energy Potential in the Contiguous United States*. Richland, Wash.: Pacific Northwest Laboratory.

Ellis, Alexander. 1991. Vice President, US Windpower. Personal communication.

Energy Information Administration (EIA). 1990. *Electric Power Plant Costs*. Washington, D.C.

———. 1991a. *Annual Energy Outlook 1991*. Washington, D.C.

———. 1991b. *Annual Energy Review 1990*. Washington, D.C.

Energy Performance Systems (EPS). 1990. "Whole Tree Burning Combustion: A Promising Technology." Minneapolis, Minn.

English, Helen. 1992. Passive Solar Industries Council. Personal communication.

Environmental Defense Fund (EDF). 1985. *To Burn or Not to Burn: The Economic Advantages of Recycling Over Garbage Incineration for New York City*. New York.

Environmental Protection Agency (EPA). 1988. *The Potential Effects of Global Climate Change on the United States*. Draft Report to Congress. Washington, D.C.

———. 1990. *Policy Options for Stabilizing Global Climate*. Report to Congress. Washington, D.C.

Etkind, Norman, Norman Hudson, and Stuart Slote. 1991. "How and Why Three Vermont Public Schools Converted to Wood Chip-fired Heating Systems." In *Energy from Biomass and Wastes XIV*. Edited by Donald L. Klass. Chicago, Ill.: Institute of Gas Technology.

Federal Energy Regulatory Commission (FERC). 1988. *Hydroelectric Power Resources of the United States*. Washington, D.C.

———. 1991. Office of Public Information.

Finger, John. 1991. Magma Program, Los Alamos National Laboratory. Personal communication.

Finneran, Kevin. 1986. "Bioenergy: Anaerobic Digestion." In *Energy Innovation: Development and Status of Renewable Energy Technologies*. Washington, D.C.: Solar Energy Industries Association.

Fitzpatrick, Nigel P., Vera Hron, and Peter J. Hryb. 1991. "Aluminum in the Marine Environment: Five Years of Testing at Keahole Point, Hawaii." *Proceedings of the IEEE Oceanic Engineering Society Conference*

(Oceans '91), vol. 1. New York: Institute of Electrical and Electronics Engineers.

Flaim, Theresa, and Susan Hock. 1984. *Wind Energy Systems for Electric Utilities: A Synthesis of Value Studies*. Golden, Colo.: National Renewable Energy Laboratory.

Flavin, Christopher, and Nicholas Lenssen. 1990. "Designing a Sustainable Energy System." *State of the World 1991*. New York: Norton.

Flynn, Barbara. 1991. National Hydropower Association. Personal communication.

Fortuna, Raymond, and Allan Jelacic. 1989. "The Geopressured-Geothermal Research Program: An Overview." Natural Gas R&D Contractors Review Meeting. Morgantown, W.Va.: Morgantown Energy Technology Center.

Garfinkel, Simson. 1990. "Seeking new ways to capture the sun." *Boston Globe*, November 5.

Geller, Howard S. 1989. "National Energy-Efficiency Platform: Description and Potential Impacts." *Energy-Efficiency Issues Paper No. 2*. Washington, D.C.: American Council for an Energy-Efficient Economy.

General Electric (GE). 1977. "Wind Energy Mission Analysis." COO/2578-1/1. Philadelphia: General Electric Space Division.

Geothermal Resources Research Conference (GRRC). 1972. *Geothermal Energy: A National Proposal for Geothermal Resources Research*. Sponsored by the National Science Foundation. Washington, D.C.: Government Printing Office.

Giese, R.F., and J.D. Rogers. 1989. "Superconducting Magnetic Energy Storage." *Applied Superconductivity*, chapter 7. Park Ridge, N.J.: Noyes Data Corporation.

Gipe, Paul A. 1991. "Wind Energy Comes of Age." *Energy Policy* (October): 756–768.

———. 1992. Paul Gipe and Associates. Personal communication.

Goldemberg, Jose, Thomas B. Johansson, Amulya K.N. Reddy, and Robert H. Williams. 1987. *Energy for a Sustainable World*. Washington, D.C.: World Resources Institute.

Gordon, Deborah. 1991. *Steering a New Course: Transportation, Energy, and the Environment*. Washington, D.C.: Island Press.

Gormley, John H. 1988. "A New Kind of Water Power." *Baltimore Sun*, August 7.

Grubb, Michael J. 1988. "The Wind of Change." *New Scientist* (March 17), 43–46.

———. 1990. "The Cinderella Options: A Study of Modernized Renewable Energy Technologies: Part 2 — A Policy Assessment." *Energy Policy* (October), 525–542.

Hagerman, George, and Ted Heller. 1988a. "Wave Energy: A Survey of Twelve Near-Term Technologies." Presented at the International Renewable Energy Conference, Honolulu, Hawaii. Alexandria, Va.: SEASUN Power Systems.

———. 1988b. "Wave Energy Technology Assessment for Grid-Connected Utility Applications." Alexandria, Va.: SEASUN Power Systems.

Haines, Eldon. 1991. Sage Advance. Personal communication.

Hall, Charles A.S., Cutler J. Cleveland, and Robert Kaufmann. 1986. *Energy and Resource Quality: The Ecology of the Economic Process*. New York: J. Wiley and Sons.

Hansen, Kent, Dietmar Winje, Eric Beckjord, Elias P. Gyftopoulos, Michael Golay, and Richard Lester. 1989. "Making Nuclear Power Work: Lessons from Around the World." *Technology Review* 92, no. 2 (February/March): 30–40.

Harris Poll. 1989. "Favor or Oppose the Building of More Nuclear Power Plants?" Conducted by telephone with 1,248 adults nationwide, December 2–6, 1988. Los Angeles: Creators Syndicate.

Harza Engineering Company. 1990. "Pumped-Storage Planning and Evaluation Guide." Palo Alto, Calif.: Electric Power Research Institute.

Helgeson, Paul. 1992. Minnesota Department of Public Service. Personal communication.

Hillesland, T., Jr. 1989. "Summary: Central Receiver Utility Study Activities." *Proceedings of the Annual Solar Thermal Technology Research and Development Conference*. SAND89-0463. Albuquerque, N.Mex.: Sandia National Laboratories.

Hinman, Norman D. 1990. "Ethanol Production from Cellulosic Biomass." Golden, Colo.: National Renewable Energy Laboratory.

Hock, Susan M., Robert W. Thresher, and Joseph M. Cohen. 1990. "Performance and Cost Projections for Advanced Wind Turbines."

Presented at the American Society of Mechanical Engineers Winter Annual Meeting. Golden, Colo.: National Renewable Energy Laboratory.

Holdren, John P., Kent B. Anderson, Peter M. Deibler, Peter H. Gleuick, Irving M. Mintzer, and Gregory P. Morris. 1983. "Health and Safety Impacts of Renewable, Geothermal, and Fusion Energy." In *Health Risks of Energy Technologies*. Edited by Curtis C. Travis and Elizabeth L. Etnier. Boulder, Colo.: Westview Press.

Houghton, Richard A., and George M. Woodwell. 1989. "Global Climatic Change." *Scientific American* 260, no. 4 (April): 36–44.

Hulstrom, Roland L., ed. 1989. *Solar Resources*. Cambridge, Mass.: MIT Press.

Huttrer, Gerald W. 1990. "Geothermal Electric Power — A 1990 World Status Update." Geothermal Resources Council *Bulletin* 19, no. 7 (July / August).

Intergovernmental Panel on Climate Change (IPCC). 1990. J.T. Houghton, G.J. Jenkins, and J.J. Ephraums, *Climate Change: The IPCC Scientific Assessment*. New York: Cambridge University Press.

International Energy Agency (IEA). 1987. *Renewable Sources of Energy*. Washington, D.C.: Organization for Economic Cooperation and Development.

———. 1989. "Conclusions and Recommendations from Task VII: Central Solar Heating Plants with Seasonal Storage." Washington, D.C.: Organization for Economic Cooperation and Development.

———. 1991. *Energy Policies and Programmes of IEA Countries: 1990 Review*. Washington, D.C.: Organization for Economic Cooperation and Development.

Johnson, F.A., 1991. "Small Land Based Ocean Thermal Energy Plants." *Proceedings of the IEEE Oceanic Engineering Society Conference (Oceans '91)*, vol. 1. New York: Institute of Electrical and Electronics Engineers.

Jones, Gary. 1992. Sandia National Laboratory. Personal communication.

Kearney, D., H. Price, I. Replogle, T. Manes, J. Costanzo, Y. Gilon, and S. Walzer. 1991. "Status of the SEGS Plants." In *1991 Solar World Congress Proceedings*, vol. 1, part 1. Edited by M.E. Arden, Susan M.A. Burley, and Martha Coleman. New York: Pergamon Press.

Kendall, Henry W. 1991. "The Failure of Nuclear Power." In *Risk, Organizations, and Society*. Edited by Martin Shubik. Boston: Kluwer Academic Publishers.

Kendall, Henry W., and Steven J. Nadis, eds. 1980. *Energy Strategies: Toward a Solar Future*. Cambridge, Mass.: Ballinger.

Kerr, Richard A. 1991. "Geothermal Tragedy of the Commons." *Science* 253: 134–135 (July 12).

Kiser, Jonathan. 1991. "Municipal Waste Combustion in the United States: An Overview." *Waste Age* (November): 27–29.

Klass, Donald L. 1988. "The U.S. Biofuels Industry." Report presented at the International Renewable Energy Conference, September 18–24. Chicago: Institute of Gas Technology.

Kolb, James O., and Kenneth E. Wilkes. 1990. "Power Generation from Waste Incineration." *Energy Engineering* 87, no. 3: 49–62.

Komanoff, Charles. 1981. *Power Plant Cost Escalation*. New York: Komanoff Energy Associates.

Larson, E.D. 1990. "Biomass-Gasifier/Gas Turbine Applications in the Pulp and Paper Industry." Prepared for the Ninth EPRI Conference of Coal Gasification Power Plants. Princeton, N.J.: Center for Energy and Environmental Studies, Princeton University.

Larson, E.D., and R.H. Williams. 1990. "Biomass-Gasifier Steam-Injected Gas Turbine Cogeneration." *Journal of Engineering for Gas Turbines and Power* 112 (April): 157–163.

Levine, Jules D., Gregory B. Hotchkiss, and Milfred D. Hammerbacher. 1991. "Basic Properties of the Spheral Solar Cell." *Proceedings of the 22nd IEEE Photovoltaic Specialists Conference*. New York: Institute of Electrical and Electronics Engineers.

Lienau, Paul J. 1991. Director, Geo-Heat Center, Oregon Institute of Technology. Personal communication.

Lippmann, M.J., and G.S. Bodvarsson. 1990. "Reservoir Technology Research at LBL Addressing Geysers Issues." *Proceedings: The National Energy Strategy — The Role of Geothermal Technology Development*. CONF-9004131. Springfield, Va.: National Technical Information Service.

Lipschutz, Ronnie D. 1980. *Radioactive Waste: Politics, Technology, and Risk*. Cambridge, Mass.: Ballinger.

Lotker, Michael. 1991. *Barriers to Commercialization of Large-Scale Solar Electricity: Lessons Learned from the LUZ Experience*. SAND91-7014. Albuquerque, N.Mex.: Sandia National Laboratory.

Lovins, Amory B., and L. Hunter Lovins. 1991. "Least-Cost Climatic Stabilization." *Annual Review of Energy and the Environment* 16: 433–531.

Luft, W., B. Stafford, and B. von Roedern. 1991. "The Photovoltaic Amorphous Silicon Research Program in the United States." In *1991 Solar World Congress Proceedings*, vol. 1, part 1. Edited by M.E. Arden, Susan M.A. Burley, and Martha Coleman. New York: Pergamon Press.

Lund, John W. 1988. "Geothermal Heat Pumps: Utilization in the United States." *Geo-Heat Center Quarterly Bulletin* 11, no. 1 (Summer).

Lund, John W., Paul J. Lienau, and G. Gene Culver. 1990. "The Current Status of Geothermal Direct Use Development in the United States — Update: 1985–1990." Geothermal Resources Council *Transactions* 14, part 1 (August).

Lunis, Ben C. 1990. "Geopressured-Geothermal Direct Use Developments." Geothermal Resources Council *Transactions* 14, part 1 (August).

Lynd, Lee R. 1989. "Large-Scale Fuel Ethanol from Lignocellulose: Potential, Economics, and Research Priorities." Dartmouth, N.H.: Thayer School of Engineering, Dartmouth College.

Lynd, Lee R., Janet H. Cushman, Roberta J. Nichols, and Charles E. Wyman. 1991. "Fuel Ethanol from Cellulosic Biomass." *Science* 251: 1318–1323 (March 15).

MacKenzie, James J., and Mohamed T. El-Ashry. 1988. *Ill Winds: Airborne Pollution's Toll on Trees and Crops*. Washington, D.C.: World Resources Institute.

Manne, A.S., and R.G. Richels. 1989. "CO_2 Emissions Limits: An Economic Analysis for the U.S.A." *Energy Journal* 10, no. 11 (November).

Maycock, Paul D. 1992a. "World PV Module Shipments." *PV News* 11, no. 2 (February): 1.

———. 1992b. "PVUSA to Rebid Kerman Substation at 500 kW Versus 200 kW." *PV News* 11, no. 2 (February): 7.

McCormick, Michael. 1981. *Ocean Wave Energy Conversion*. New York: J. Wiley and Sons.

McGowan, Jon G. 1991. "Large-Scale Solar/Wind Electrical Production Systems — Predictions for the 21st Century." In *Energy and the Environment in the 21st Century*. Edited by Jefferson W. Tester, David O. Wood, and Nancy A. Ferrari. Cambridge, Mass.: MIT Press.

McKay, H.G., R.E. Steffens, and B.W. Curlin. 1989. "Review of the Design Methodology for the Bad Creek Underground Powerhouse As It Would Apply to a Hard Rock Cavern Design for Compressed Air Energy Storage (CAES)." In *Storage of Gases in Rock Caverns*. Edited by Bjorn Nilsen and Jorn Olsen. Brookfield, Vt.: Gower Publishing.

McKinley, Kelton R., Sidney H. Browne, D. Richard Neill, Arthur Seki, and Patrick K. Takahashi. 1990. "Hydrogen Fuel From Renewable Resources." *Energy Sources* 12.

McLarnon, Frank R., and Elton J. Cairns. 1989. "Energy Storage." *Annual Review of Energy* 14: 241–271.

Meridian Corporation. 1989. "Energy Systems Emissions and Materiel Requirements." Report prepared for the Department of Energy, Office of Renewable Energy. Washington, D.C.: Department of Energy.

Metz, W.D., and A.L. Hammond. 1978. *Solar Energy in America*. Washington, D.C.: American Association for the Advancement of Science.

MHB Technical Associates. 1990. *Advanced Reactor Study*. Cambridge, Mass.: Union of Concerned Scientists.

Miles, Thomas R., and Thomas R. Miles, Jr. 1991. "Urban Wood: Fuel from Landscapers and Landfills." *Biologue* 8, no. 3 (September): 10–12.

Mock, John E., and Gene V. Beeland. 1988. "Geothermal Energy — A Secure, Cost-Competitive Energy Source for the U.S." *Energy Technology XV: Repowering America*. Rockville, Md.: Government Institutes.

Moskovitz, David. 1991. "Renewable Energy: Regulatory Barriers and Opportunities." Washington, D.C.: World Resources Institute. Draft.

National Academy of Sciences (NAS). 1987. *Geothermal Energy Technology: Issues, R&D Needs, and Cooperative Arrangements*. Washington, D.C.: National Academy Press.

———. 1991. *Policy Implications of Greenhouse Warming*. Washington, D.C.: National Academy Press.

National Energy Strategy (NES). 1991. *National Energy Strategy: Powerful Ideas for America* (First Edition). Arlington, Va.: National Technical Information Service.

National Renewable Energy Laboratory (NREL). 1990a. *Insolation Data Manual and Direct Normal Solar Radiation Data Manual*. Golden, Colo.

———. 1990b. *The Potential of Renewable Energy: An Interlaboratory White Paper*. Golden, Colo.

National Science Foundation (NSF). 1972. *An Assessment of Solar Energy as a National Resource*. Report of the NSF/NASA Solar Energy Panel. Washington, D.C.

Negus-deWys, J. 1990. "The Geopressured-Geothermal Resource: Research and Use." *Proceedings: The National Energy Strategy — The Role of Geothermal Technology Development*. CONF-9004131. Springfield, Va.: National Technical Information Service.

———. 1991. Idaho National Engineering Laboratory. Personal communication.

———. 1992. Idaho National Engineering Laboratory. Personal communication.

New England River Basins Commision. 1981. *Water, Watts, and Wilds: Hydropower and Competing Uses in New England*. Cambridge, Mass.: Union of Concerned Scientists.

Newsday. 1989. *Rush to Burn: Solving America's Garbage Crisis?* Washington, D.C.: Island Press.

Nicholson, Robert J. 1990. Vice President, Sea Solar Power. Personal communication.

Nordhaus, William. 1989. *The Economics of the Greenhouse Effect*. New Haven, Conn.: Yale University.

Office of Technology Assessment (OTA). 1989. *Facing America's Trash: What Next for Municipal Solid Waste?* Washington, D.C.

Ogden, Joan M., and Robert H. Williams. 1989. *Solar Hydrogen: Moving Beyond Fossil Fuels*. Washington, D.C.: World Resources Institute.

Ogden, Joan M., Robert H. Williams, and Mark E. Fulmer. 1991. "Cogeneration Applications of Biomass Gasifier/Gas Turbine Technologies in the Cane Sugar and Alcohol Industries." In *Energy and the Environment in the 21st Century*. Edited by Jefferson W. Tester, David O. Wood, and Nancy A. Ferrari. Cambridge, Mass.: MIT Press.

O'Regan, Brian, and Michael Gratzel. 1991. "A Low-Cost, High-Efficiency Solar Cell Based on Dye-Sensitized Colloidal TiO_2 Films." *Nature* 353, no. 6346 (October 24): 737–740.

Osborne, Don. 1992. Sacramento Municipal Utility District. Personal communication.

Ottinger, Robert L., David R. Wooley, Nicholas A. Robinson, David R. Hodas, and Susan E. Babb. 1990. *Environmental Costs of Electricity*. New York: Oceana.

Pacific Northwest Laboratory (PNL). 1986. *Wind Energy Resource Atlas of the United States*. Golden, Colo.: National Renewable Energy Laboratory.

Pasqualetti, M.J., and Mark Dellinger. 1989. "Hazardous Waste from Geothermal Energy: A Case Study." *Journal of Energy and Development* 13, no. 2 (Spring): 275–295.

Passive Solar Industries Council (PSIC). 1990. *Passive Solar Design Strategies: Guidelines for Home Builders*. Washington, D.C.

Penney, T., and D. Bharathan. 1987. "Power from the Sea." *Scientific American* 256, no. 1 (January).

Pimentel, David. 1991. "Ethanol Fuels: Energy, Security, Economics, and the Environment." *Journal of Agricultural and Environmental Ethics* 4: 1–13.

Premuzic, Eugene T., Mow S. Lin, and Sun Ki Kang. 1990. "Developments in Geothermal Waste Treatment Biotechnology." *Proceedings: The National Energy Strategy—The Role of Geothermal Technology Development*. CONF-9004131. Springfield, Va.: National Technical Information Service.

Public Power Weekly. 1991. "California Utilities Launch Solar Two; $39 Million Project to Begin 1994." *Public Power Weekly* (September 9).

Ramanathan, V. 1988. "The Greenhouse Theory of Climate Change: A Test by an Inadvertent Global Experiment." *Science* 240: 293–299 (April 15).

Rashkin, Sam. 1991. "California Wind Project Performance: A Review of Wind Performance Results from 1985 to 1990." Presented at the annual meeting of the American Wind Energy Association (Windpower '91). Sacramento, Calif.: California Energy Commission.

Reganold, John P., Robert I. Papendick, and James F. Parr. 1990. "Sustainable Agriculture." *Scientific American* 262, no. 6 (June): 112–120.

Robinson, Gordon. 1988. *The Forest and the Trees: A Guide to Excellent Forestry*. Washington, D.C.: Island Press.

Rogers, Wayne L. 1989. "Hydropower: The Forgotten Technology." Washington, D.C.: National Hydropower Association.

Russell, A.G., D. St. Pierre, and J.B. Milford. 1990. "Ozone Control and Methanol Fuel Use." *Science* 247: 201–204 (January 12).

Sanders, M.M. 1991. "Energy from the Oceans." In *The Energy Sourcebook*. Edited by Ruth Howes and Anthony Fainberg. New York: American Institute of Physics.

Sandia National Laboratories (SNL). 1987a. *Today's Photovoltaic Systems: An Evaluation of Their Performance*. SAND87-2585. Albuquerque, N.Mex.

———. 1987b. *The Interconnection Issues of Utility-Intertied Photovoltaic Systems*. SAND87-3146. Albuquerque, N.Mex.

———. 1991a. "Cummins Awarded Contract to Develop Economical Dish-Stirling Power System." News release, October 16.

———. 1991b. *Today's Solar Power Towers*. SAND91-2018. Albuquerque, N.Mex.

Saunders, David. 1991. National Association of Home Builders. Personal communication.

Schipper, L., R. Howarth, and H. Geller. 1990. "United States Energy Use from 1973 to 1987: The Impacts of Improved Efficiency." *Annual Review of Energy* 15: 455–504.

Schneider, Stephen H., and Rondi Londer. 1984. *The Coevolution of Climate and Life*. San Francisco: Sierra Club Books.

Shea, Cynthia Pollock. 1988. "Renewable Energy: Today's Contribution, Tomorrow's Promise." *Worldwatch Paper No. 81*. Washington, D.C.: Worldwatch Institute.

Shugar, Daniel S. 1991. "Photovoltaics in the Utility Distribution System: The Evaluation of System and Distributed Benefits." *PG&E PV Conference Papers 1984-1990*. 007.3-91.2. San Ramon, Calif.: Pacific Gas and Electric, Research and Development Division.

Sissine, Fred J. 1992. "Renewable Energy: A New National Commitment?" CRS Issue Brief IB90110. Washington, D.C.: Congressional Research Service.

Smith, D.R. 1990. "Wind Energy Resource Potential and the Hourly Fit of Wind Energy to Utility Loads in Northern California." Presented at the annual meeting of the American Wind Energy Association (Windpower '90). San Ramon, Calif.: Pacific Gas and Electric, Research and Development Division.

Smith, D.R., and M.A. Ilyin. 1991. "Wind and Solar Energy, Costs and Value." Presented at the Tenth American Society of Mechanical Engineers Wind Energy Symposium. San Ramon, Calif.: Pacific Gas and Electric, Research and Development Division.

Spanner, G.E., and G. L. Wilfert. 1989. "Potential Industrial Applications for Composite Phase-Change Materials As Thermal Energy Storage Media." PNL-6935. Richland, Wash.: Pacific Northwest Laboratory.

Spencer, D.F. 1991. "A Preliminary Assessment of Carbon Dioxide Mitigation Options." *Annual Review of Energy and the Environment* 16: 259–73.

Sperling, Daniel. 1988. *New Transportation Fuels*. Los Angeles: University of California.

Stevens, William K. 1992. "New Studies Predict Profits in Heading Off Warming." *New York Times*. March 17.

Sunderland, J. Edward. 1991. CSHPSS Development Laboratory, University of Massachusetts, Amherst. Personal communication.

Sunderland, J. Edward, and Dwayne S. Breger. 1990. "The Development of a Central Solar Heating Plant with Seasonal Storage at the University of Massachusetts/Amherst." Amherst, Mass.: CSHPSS Development Laboratory, University of Massachusetts.

Takahashi, Patrick, D. Richard Neill, Victor D. Phillips, and Charles M. Kinoshita. 1990. "Hawaii: An International Model for Methanol from Biomass." *Energy Sources* 12: 421–428.

Taschini, A., and J.J. Iannucci. 1991. "Potential of Photovoltaic Systems for Present and Future Electric Utility Applications." *PG&E PV Conference Papers 1984-1990*. 007.3-91.2. San Ramon, Calif.: Pacific Gas and Electric, Research and Development Division.

Tester, Jefferson W., D.W. Brown, and R.M. Potter. 1989. *Hot Dry Rock Geothermal Energy — A New Energy Agenda for the 21st Century*. LA-11514-MS. Los Alamos, N.Mex.: Los Alamos National Laboratory.

Tester, Jefferson W., and Howard J. Herzog. 1990. *Economic Predictions for Heat Mining: A Review and Analysis of HDR Geothermal Energy Technology*. MIT-EL 90-001. Cambridge, Mass.: MIT Energy Laboratory.

Thayer, Robert L., and Heather A. Hansen. 1991. *Wind Farm Siting Conflicts in California: Implications for Energy Policy*. Center for Design Research, University of California, Davis.

Tomlinson, J., and L. Kannberg. 1990. "Thermal Energy Storage." *Mechanical Engineering* 112, no. 9 (September): 68–72.

Trexler, Dennis T., Tomas Flynn, and James L. Hendrix. 1990. "Heap Leaching." *Geo-Heat Center Quarterly Bulletin* 12, no. 4 (Summer 1990).

Tyson, K. Shaine. 1990. *Biomass Resource Potential of the United States*. Golden, Colo.: National Renewable Energy Laboratory.

———. 1991. *Resource Assessment of Waste Feedstocks for Energy Use in the Western Regional Biomass Energy Area*. WAPA Project BF983232. Golden, Colo.: National Renewable Energy Laboratory.

Ullal, Harin S., Kenneth Zweibel, Richard L. Mitchell, and Rommel Noufi. 1991. "Polycrystalline Thin Film Photovoltaic Technology." In *1991 Solar World Congress Proceedings*, vol. 1, part 1. Edited by M.E. Arden, Susan M.A. Burley, and Martha Coleman. New York: Pergamon Press.

Union of Concerned Scientists (UCS), Alliance to Save Energy, American Council for an Energy-Efficient Economy, and Natural Resources Defense Council. 1991. *America's Energy Choices: Investing in a Strong Economy and a Clean Environment*. Cambridge, Mass.: Union of Concerned Scientists.

United States Geological Survey (USGS). 1979. *Assessment of Geothermal Resources of the United States*. Circular 790. Edited by Patrick L.J. Muffler. Washington, D.C.: U.S. Geological Survey.

——— 1983. *Assessment of Low-Temperature Geothermal Resources of the United States — 1982*. Circular 892. Edited by Marshall J. Reed. Washington, D.C.: U.S. Geological Survey.

U.S. Advanced Battery Consortium. 1991. Press release. October 25.

U.S. Department of Agriculture (USDA). 1991a. "Michigan Timber Industry — An Assessment of Timber Product Output and Use, 1988." NC-121. St. Paul, Minn.: North Central Forest Experiment Station.

———. 1991b. "Minnesota Timber Industry — An Assessment of Timber Product Output and Use, 1988." NC-127. St. Paul, Minn.: North Central Forest Experiment Station.

———. 1991c. "Wisconsin Timber Industry — An Assessment of Timber Product Output and Use, 1988." NC-124. St. Paul, Minn.: North Central Forest Experiment Station.

———. 1991d. "Assessing Removals for North Central Forest Inventories." NC-299. St. Paul, Minn.: North Central Forest Experiment Station.

Von KleinSmid, William. 1991. Southern California Edison. Personal communication.

Wald, Matthew. 1991. "A Tough Sell for Electric Cars." *New York Times*, November 26.

Weinberg, C.J., S.L. Hester, and T.U. Townsend. 1991. "The Photovoltaics for Utility Scale Applications (PVUSA) Project." *PG&E PV Conference Papers 1984–1990*. 007.3-91.2. San Ramon, Calif.: Pacific Gas and Electric, Research and Development Division.

Weinberg, Carl J., and Robert H. Williams. 1990. "Energy from the Sun." *Scientific American* 263, no. 3 (September).

Wendland, Ronald D. 1990. "Commercial Cool Storage." *Energy Engineering* 87, no. 6 (June): 18–22.

Wick, G.L., and W.R. Schmitt. 1981. *Harvesting Ocean Energy*. Paris: UNESCO Press.

Williams, Carl. 1991. LaJet Incorporated. Personal communication.

Williams, R.H. 1990. "Low-Cost Strategies for Coping with CO_2 Emission Limits." *Energy Journal* 11, no. 3 (March).

World Resources Institute (WRI). 1990. *World Resources 1990–91*. New York: Basic Books.

Wyman, Charles E., and Norman D. Hinman. 1990. "Ethanol: Fundamentals of Production from Renewable Feedstocks and Use As a Transportation Fuel." *Applied Biochemistry and Biotechnology* 24/25: 735–753.

Index

Air pollution, 6, 11
  associated with biomass, 106-108
  costs of, 28-29
Alternative energy strategies, 34-37
Anaerobic digestion of biomass, 104-105

Batteries, 161-163
Biogas, 104-105
Biomass, 87-110
  biochemical conversion of, 101-105
  direct combustion of, 95-97
  electricity generation from, 96, 99-100, 104-105
  energy balance of, 104
  energy crops, 92-95
  environmental issues with, 105-109, 184
  forests, 91, 93, 109
  history of, 87-88, 97-98, 101-102
  implications for agriculture and forestry, 108-109
  land requirements for, 92-94
  municipal solid waste, 91, 97, 107, 184
  phenols from, 100
  resources of, 88-95
  thermochemical conversion of, 97-101

CAES. *See* Compressed air energy storage
CAFE standards. *See* Corporate Average Fuel Economy standards
Carbon-dioxide emissions, 7-12, 35-36, 108
Central receivers, 55-57
CFCs. *See* Chlorofluorocarbons
Chlorofluorocarbons, 6, 7, 10, 11, 20n
Clean Air Act, 11, 32
Climate change. *See* Global warming
Compressed air energy storage, 165-167
Corporate Average Fuel Economy standards, 14, 15

Department of Energy
  and batteries for electric vehicles, 162
  and biomass, 92, 94, 99, 103, 105
  and energy storage, 157, 158
  and geothermal energy, 140, 150
  and the National Energy Strategy, 185n
  and ocean thermal energy conversion, 122, 123
  recommendations for R&D funding by, 182
  and solar energy, 53, 54, 55, 60, 62, 64, 157
Discount rate, 30-33
  choice of, 30-31, 36
  effect of risks on, 31-33
  importance of for evaluating renewable energy, 30
Dish-Stirling systems, 54-55

Electric Power Research Institute, 78, 96, 162, 165, 168
Electric vehicles, 161-162, 163
Energy consumption
  relation to economic output, 13-14
  U.S., 13, 23
Energy-economic models, 14-15
Energy efficiency, 13-16, 175-176
Energy policy, 12-19, 30-33, 174-185
Energy storage, 155-172
  with conventional energy systems, 25-26, 155-156
  electricity storage, 161-169
  environmental impacts of, 164, 167, 169
  hydrogen, 169-172
  thermal energy storage, 156-160
  and wind energy, 82-84
Energy supply R&D funding, 38n4
Environmental impacts
  of biomass, 105-109, 184
  of energy storage, 164, 167, 169
  of fossil fuels, 6, 10-11, 28-30

of geothermal energy, 150-153
of hydropower, 115-118
of ocean thermal energy conversion, 123-124
of solar energy, 66-67
of tidal power, 125
of wind energy, 84-85
Environmental Protection Agency, 9, 37, 107
EPA. *See* Environmental Protection Agency
EPRI. *See* Electric Power Research Institute
Ethanol, 101-104, 110n9
and air pollution, 107

Farms, energy, 91-95, 108-109
Fermentation of biomass, 101-102, 103-104
Fossil fuels
consumption of, 3, 5, 6, 11
environmental and social costs of, 6, 10-11, 28-30

Gasification of biomass, 98-100
Geothermal energy, 127-154
costs of, 134, 140, 145
direct uses of, 136-138
electricity generation from, 134-136
environmental issues with, 150-153
geopressured brines, 138-141
history of, 127-128
hot dry rock, 141-148
hydrothermal fluids (reservoirs), 131-138
magma, 148-150
nonrenewable nature of, 129-131
resources of, 127-131
siting conflict in Hawaii, 153
Geysers, The, 127, 129, 131-132, 134, 135-136, 151, 152, 153
problems at, 135-136
Global warming, 7-11, 13
causes of, 7
and observed temperature increases, 8
potential impacts of, 9-10
predictions of, 8-9
role of fossil fuels in, 10-11
U.S. role in, 11
Greenhouse effect. *See* Global warming
Greenhouse gases, 7-11, 108

HDR. *See* Hot dry rock
Heat storage
latent, 158-159
seasonal, 159-160
sensible, 157-158
thermochemical, 159

Hot dry rock, 141-148
costs of, 145, 147
resources of, 141-143
Hybrid fossil-renewable energy systems, 26, 155-156
Hydrogen
from biomass, 171
from photovoltaics, 68n3, 170-171
as a transportation fuel, 171-172
Hydropower, 111-118
environmental and regulatory issues with, 115-118, 184
history of, 111-112
resources of, 113-114
technology, 114-115

Incentives for renewable energy, 178-179

Least-cost planning, 179
Liquefaction of biomass, 101
LUZ International, Ltd., 51-53, 54, 56, 66, 68n9, 155, 158

Magma, 148-150
Methanol, 97-101, 110n8
and air pollution, 107

National Energy Strategy, 34, 116, 117-118, 173-174, 184, 185n
National Renewable Energy Laboratory, 37n1, 183
Natural gas
as backup for intermittent renewable sources, 26, 155-156
switching to, 12
NREL. *See* National Renewable Energy Laboratory
Nuclear power, 16-19
advanced designs, 18
public opinion toward, 16-17
rising cost of, 17
and waste, 18-19
and weapons proliferation, 19

Ocean thermal energy conversion, 118-124
environmental issues with, 123-124
technology and cost, 120-123
Oil crisis, next, 5-6
OTEC. *See* Ocean thermal energy conversion
Ozone layer, destruction of, 6-7

Pacific Gas and Electric, 56, 64, 65, 78, 81-82
Parabolic dishes, 54-55
Parabolic troughs, 50-53
for electric applications, 52-53

for thermal applications, 50-51
Passive solar design, 44-47
PG&E. *See* Pacific Gas and Electric
Photovoltaics, 57-66, 170-171, 182-183
  costs of, 61-62, 64
  for home use, 61
  markets for, 63-66
  polycrystalline cells, 59-60, 63
  single-crystal silicon cells, 58-59, 63
  thin films, 60
Public Utilities Regulatory Policy Act, 52, 53, 116, 134, 180
Pumped hydroelectric storage, 114, 163-164
PURPA. *See* Public Utilities Regulatory Policy Act
Pyrolysis of biomass, 100

Renewable energy
  advantages of, 22
  current use of, 23
  definition of, 1
  environmental regulation of, 183-184
  financial incentives for developing, 178-179
  history of, 21-22
  international expenditures on, 174
  market for, 26-30
  market barriers against, 26-27
  policies to promote, 176-183
  potential of, 33-37
  research and development on, 27-28, 182-183
  resources of, 23
  tax credits for, 178
  technological progress on, 22-26
  use with other energy systems, 25-26, 155-156
  U.S. R&D funding for, 21-22, 27-28, 29, 37n1, 182-183, 190-191

SERI. *See* National Renewable Energy Laboratory
SMES. *See* Superconducting magnetic energy storage
Solar collectors, 47-50
Solar energy, 39-69
  environmental issues with, 66-67
  history of, 39-40
  photovoltaic cells, 57-66
  resource of, 40-43
  solar buildings, 43-50
  solar collectors, 47-50
  solar-thermal concentrating systems, 50-57
  use of land for, 43, 66
Solar Energy Research Institute. *See* National Renewable Energy Laboratory
Solar One, 55, 56
Solar power. *See* Solar energy
Solar-thermal concentrating systems, 50-57
Solar Two, 55-56
Southern California Edison, 52, 59, 159, 163
Storage. *See* Energy storage; Heat storage
Superconducting magnetic energy storage, 167-169
Syngas, 97-101

Tax bias against renewables, 27, 53
Tax credits, proposed for renewable energy, 178
Taxes, proposed on fossil fuels, 176-177
Tidal power, 124-125
Time value of money. *See* Discount rate

US Windpower, 78-79, 80, 83
Utility regulation, proposed changes in, 179-180

Wastes, for energy production, 89-91
Wave power, 125-126
Wind energy, 71-85
  environmental issues with, 84-85
  history of, 72-73
  intermittency, 75, 81-84
  public attitudes toward, 83
  resource of, 72-75
  technology and cost, 75-84
Wind power. *See* Wind energy
Wind turbines, 24, 75-81
  intermediate-size, 75-80
  small, 80-81

98 16